McGRAW-HILL RYERSON
MATHEMATICS 7
MAKING CONNECTIONS
Student Workbook

ADVISORS

Dan Antflyck
Toronto District School Board

L. G. Ban
Ottawa-Carleton Catholic District School Board

Richard Chaplinsky
Ottawa-Carleton Catholic District School Board

Warren Dixon
District School Board of Niagara

Christina Maschas-Hammond
Peterborough, Victoria, Northumberland, Clarington Catholic District School Board

Roland W. Meisel
Toronto, Ontario

Christopher Perry
Hamilton Wentworth District School Board

D.M. Quigley
Ottawa, Ontario

Toronto   Montréal   Boston   Burr Ridge, IL   Dubuque, IA   Madison, WI   New York
San Francisco   St. Louis   Bangkok   Bogotá   Caracas   Kuala Lumpur   Lisbon   London
Madrid   Mexico City   Milan   New Delhi   Santiago   Seoul   Singapore   Sydney   Taipei

COPIES OF THIS BOOK
MAY BE OBTAINED BY
CONTACTING:

McGraw-Hill Ryerson Ltd.

WEB SITE:
http://www.mcgrawhill.ca

E-MAIL:
orders@mcgrawhill.ca

TOLL-FREE FAX:
1-800-463-5885

TOLL-FREE CALL:
1-800-565-5758

OR BY MAILING YOUR
ORDER TO:
McGraw-Hill Ryerson
Order Department
300 Water Street
Whitby, ON L1N 9B6

Please quote the ISBN
and title when placing
your order.

Student Workbook ISBN:
0-07-090952-0

McGraw-Hill Ryerson

Mathematics 7 Making Connections Student Workbook

Copyright © 2004, McGraw-Hill Ryerson Limited, a Subsidiary of The McGraw-Hill Companies. All rights reserved. No part of this publication may be reproduced or transmitted in any form or by any means, or stored in a data base or retrieval system, without the prior written permission of McGraw-Hill Ryerson Limited, or, in the case of photocopying or other reprographic copying, a licence from The Canadian Copyright Licensing Agency (Access Copyright). For an Access Copyright licence, visit www.accesscopyright.ca or call toll free to 1-800-893-5777.

ISBN 0-07-090952-0

http://www.mcgrawhill.ca

1 2 3 4 5 6 7 8 9 10 M 0 9 8 7 6 5 4

Printed and bound in Canada

Care has been taken to trace ownership of copyright material contained in this text. The publishers will gladly accept any information that will enable them to rectify any reference or credit in subsequent printings.

National Library of Canada Cataloging in Publication Data
Mathematics 7: making connections. Student Workbook.
ISBN 0-07-090952-0
1. Mathematics—Problems, exercises, etc.
QA107.2.M39 2004 Suppl.   510   C2004-902819-7

PUBLISHER: Diane Wyman
DEVELOPMENTAL EDITORS: Eileen Jung, Bradley T. Smith, Jenna Voisin, and Michael Vo of
    First Folio Resource Group, Inc.
MANAGER, EDITORIAL SERVICES: Linda Allison
SUPERVISING EDITOR: Kristi Moreau
PRODUCTION SUPERVISOR: Yolanda Pigden
PRODUCTION COORDINATOR: Janie Deneau
ELECTRONIC PAGE MAKE-UP: Tom Dart/First Folio Resource Group, Inc.

Contents

Problem Solving v

Get Ready for Grade 7 x

Chapter 1   Measurement and Number Sense 1

Chapter 2   Two-Dimensional Geometry 15

Chapter 3   Fraction Operations 23

Chapter 4   Probability and Number Sense 39

Chapter 5   Fractions, Decimals, and Percents 51

Chapter 6   Patterning 59

Chapter 7   Exponents 65

Chapter 8   Three-Dimensional Geometry and Measurement 75

Chapter 9   Data Management: Collection and Display 87

Chapter 10 Data Management: Analysis and Evaluation 97

Chapter 11 Integers 105

Chapter 12 Patterning and Equations 117

Chapter 13 Geometry of Transformations 125

Preparing for Grade 8 133

Name: _____   Date: _____

Problem Solving

Understand

Read the problem carefully.
- What information do you have?
- What are you asked to do?
- Explain the problem in your own words.

Plan

Choose one or more strategies to solve the problem.

- Make a model
- Make an assumption
- Make a picture or diagram
- Find needed information
- Choose a formula
- Solve a simpler diagram

- Act it out
- Make an organized list or tree diagram
- Work backward
- Make a table or chart
- Use systematic trial
- Look for a pattern

- Use tools such as a ruler, calculator, protractor, or number line.
- Use materials such as graph paper.
- Does this problem resemble another problem you've solved before? Use a similar strategy to solve the new problem.

Do It!

Solve the problem. Carry out your plan.
- Use mental math to estimate.
- Do the calculations.
- Record each step.
- Explain and justify your thinking.

Look Back

Examine your answer. Does it make sense?
- Is your answer reasonable given the information in the problem?
- Is your answer close to your estimate?
- Use another strategy to check your answer.
- Compare your method with that of other students.

Four problems are presented.

Twelve strategies are used to solve the problems.

Complete the solutions for each problem by completing diagrams and filling in the blanks.

After you have completed the problems, try using different strategies.

Study Skills

Create your own problems. Exchange them with a study partner. Compare the strategies you and your partner used to solve the problem. Which strategy was easier to use? Why? Write notes about which strategy may be more efficient to use.

Problem 1

Store-It! has shipping boxes that are 2 m wide and 1 m high. Only the lengths are different. Jackie has 6 m³ of material to ship. What length of box does she need?

Make a model

Use sugar cubes.
One sugar cube represents 1 m³.

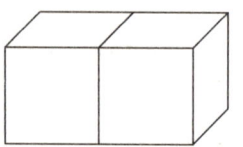

a) Complete the diagram of the model.

b) Add the measure to the diagram.

Hint
The volume of the two cubes is 2 m³.
How many of these models will fit into 6 m³?

c) The length of the box is _____.

Make a picture or diagram

Use graph paper to make a diagram of the box.

a) Complete the diagram of the box.

b) Add labels to the diagram.

c) The length of the box is _____.

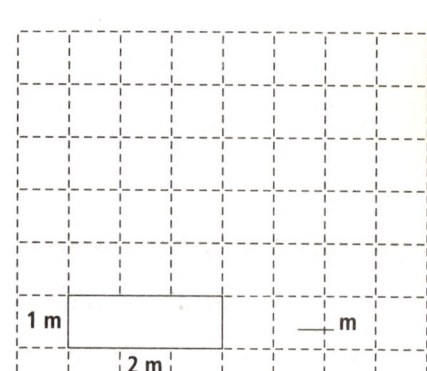

Choose a formula

The formula for the volume, V, of a rectangular prism is

$V = l \times w \times h$

Substitute $V = 6$, $w = 2$, and $h = 1$.

____ = l × ____ × ____

____ = l × ____

____ = ____ × ____

The length of the box is _____.

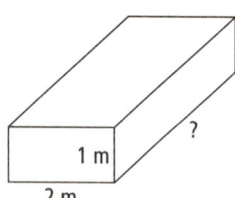

Study Skills

For some problems, you could use more than one problem-solving strategy.
In this problem, the Choose a formula strategy is used. A diagram is also included to help visualize the problem.

Name: _____ Date: _____

Problem 2

Yasmin wants to buy CDs with $40 she received for her birthday. How many CDs can she buy, if there is no tax?

Make an assumption

a) Assume that the average cost of one CD is _____.

Hint
The price of a CD at CD Master is from $17 to $21 each.

Find the number of CDs she can buy.

b) Show your work.

Yasmin can buy _____ CDs.

Find needed information

You notice an advertisement on television for CD Master stores. The store is having a special of $9.99 per CD, tax included. Find the number of CDs Yasmin can buy. Show your work.

Yasmin can buy _____ CDs.

Name: _____   Date: _____

Problem 3

Mrs. Mason's class is having a party. She collects the same amount of money from each of her 25 students. Then she adds $10 for a total of $85. How much did each student give?

Act it out

Divide by 25 students.

a) Mrs. Mason gave _____.

b) The students gave _____.

c) There are _____ students.

d) Which mathematical operation should I use to find the amount each student gave? Explain your choice.

e) This means each student gave _____.

Work backward

a) Mrs. Mason gave _____.

b) 85 – _____ = _____

c) There are 25 students.

d) Divide _____ by 25.

e) Each student gave _____.

Use systematic trial

The total amount collected is 10 + 25 × each student's amount.

a) Try $2:
   10 + 25 × ____
   = 10 + ____
   = ____ (Too low)

   Try $3:
   10 + 25 × ____
   = 10 + ____
   = ____ (_____)

   Try $4:
   10 + 25 × ____
   = 10 + ____
   = ____ (_____)

b) Each student gave _____.

Name: _____   Date: _____

Problem 4

Mei is making friendship pins. Each pin will have two coloured beads. She has five different-coloured beads: red, blue, purple, gold, and silver. How many pins can she make? (Order of colour doesn't matter and a pin can have beads of the same colour.)

Solve a simpler problem

a) If there are only two choices, say red and blue, then there are _____ possible pins.

b) The pins' possible colours are: red and _____, red and _____, blue and _____.

Now I have a starting point to finish solving the problem.

Make a picture or diagram

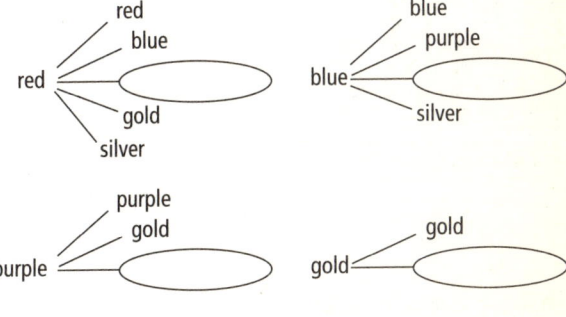

a) Fill in the circles to complete the tree diagram.

b) number of pins with different coloured beads = _____

c) number of pins with same coloured beads = _____

d) total number of pins = _____ + _____
   = _____

Make a table or chart

a) Show which bead combinations are possible.
Place an X in each cell that represents a possible combination.
Remember: order of colour doesn't matter, and a pin may have beads of the same colour.

	red	blue	purple	gold	silver
red					
blue					
purple					
gold					
Silver					

b) Count the Xs.

The number of pins possible is _____.

1 Fractions, Metric Units, Estimation

Get Ready Mentally

1. Which is greater? How do you know?

   a) $\frac{2}{5}$ or $\frac{2}{6}$ b) $\frac{9}{2}$ or $\frac{3}{2}$ c) $\frac{1}{8}$ or $\frac{1}{7}$

   d) $\frac{2}{3}$ or $\frac{3}{2}$ e) $\frac{12}{5}$ or $\frac{8}{9}$ f) $\frac{1}{3}$ or $\frac{3}{1}$

2. State whether each measurement is greater than or less than 1000 cm. How do you know?

   a) 1.5 m b) 10 m

   c) 5 m + 6 m d) 400 cm + 4 m

Get Ready by Thinking

Choose the most reasonable estimate in each question. Explain your thinking.

3. The height of your classroom blackboard is about

   A 2 m B 100 cm
   C 4 m D 1.5 m

4. The mass of this workbook is about

   A 1 kg B 3 kg
   C 500 g D 10 000 g

5. The volume of milk is about

   A 350 mL B 1 L
   C 1025 mL D 4 L

6. The distance between two soccer goalposts is about

   A 1 km B 1050 m
   C 8 m D 350 m

7. The shaded portion is about

   A $\frac{1}{4}$ B $\frac{1}{3}$
   C $\frac{1}{2}$ D $\frac{3}{4}$

   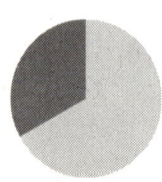

8. The time you spend in school in one day is about

   A $\frac{16}{24}$ B $\frac{7}{24}$
   C $\frac{11}{24}$ D $\frac{10}{24}$

9. The mass of a telephone is about

   A 300 g B 0.7 kg
   C 0.150 kg D 0.56 g

   | 1 cm = 10 mm |
   | 1 m = 100 cm |
   | 1 g = 1000 mg |
   | 1 kg = 1000 g |
   | 1 L = 1000 mL |

10. Refer to the chart. The cost of the camping supplies is about

    Tent $49.99
    Mosquito Repellent $ 14.49
    Sleeping Bag $32.95
    Compass $3.40

    A $70 B $100 C $89 D $130

11. The length of a small chalkboard eraser is about

    A 15 cm B 0.7 m C 90 cm D 40 cm

12. The height of your school desk is about

    A 1 m B 100 cm C 20 cm D 60 cm

13. The number of school days in a year is about

    A 200 B 300 C 275 D 330

Name: _____ Date: _____

2 Multiplying and Dividing Decimals, Estimation

Get Ready Mentally

1. Solve.
   a) $43 \times 10 =$
   b) $43 \times 1000 =$
   c) $43 \times 0.001 =$
   d) $43 \times 0.1 =$

2. Solve.
   a) $43 \div 10 =$
   b) $43 \div 1000 =$
   c) $43 \div 0.001 =$
   d) $43 \div 0.1 =$

3. $43 \times 8 = 344$
   Use this fact to find each product.
   a) $43 \times 80 =$
   b) $4.3 \times 8 =$
   c) $43 \times 0.8 =$
   d) $4.3 \times 0.8 =$

4. $4308 \div 6 = 718$
   Use this fact to find each quotient.
   a) $430.8 \div 6 =$
   b) $4308 \div 60 =$
   c) $43.08 \div 6 =$
   d) $4308 \div 600 =$

Get Ready by Thinking

Choose the most reasonable estimate for questions 5 to 10.

5. About how many grams of chocolate are there?

   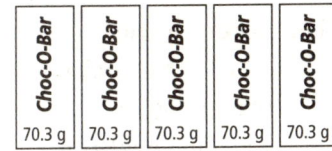

   A 355 g   B 325 g   C 350 g   D 360 g

6. Heather can run the 414-m race six times. About how far can she run?

   A 2520 m   B 2050 m
   C 2400 m   D 2600 m

7. About how many grams of biscuits did the dog eat?

   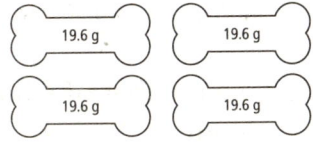

   A 80 g   B 72 g   C 76 g   D 100 g

8. There are 126 grapes on 9 slices of fruit flan. About how many grapes are on each slice?

   A 12 grapes   B 13 grapes
   C 14 grapes   D 15 grapes

9. Pablo sold 11 cups of lemonade. About how many litres of lemonade is this?

   250 mL

   A 2.0 L   B 2.5 L
   C 2.70 L   D 3.5 L

10. Opal has 348 small candies to put into treat bags for her 7 friends. About how many candies will each friend receive?

    A 35 candies   B 48 candies
    C 50 candies   D 70 candies

3 Patterns With Natural Numbers, Fractions, and Decimals

Get Ready Mentally

1. Identify the next three numbers in each pattern.

   a) 1, 4, 7, ____, ____, ____

   b) 9, 18, 27, ____, ____, ____

   c) 2, 8, 14, ____, ____, ____

   d) 1, 2, 5, 10, 17, ____, ____, ____

   e) 1, 3, 7, 13, ____, ____, ____

   f) 1, 2, 3, 5, 8, 13, ____, ____, ____

2. Identify the next three numbers in each pattern.

   a) 1.2, 1.6, 2.0, ____, ____, ____

   b) $\frac{1}{3}, \frac{1}{6}, \frac{1}{12}, \frac{1}{24},$ ____, ____, ____

   c) 4.27, 4.34, 4.41, ____, ____, ____

   d) $\frac{1}{2}, 1, \frac{3}{2},$ ____, ____, ____.

Study Skills

Work with a study partner. Create a pattern that uses multiplication and addition. Exchange patterns. Describe the pattern.

Get Ready by Thinking

3. Explain what happens to the input number to get the output number.

   a)
Input	Output
5	12
7	14
11	18

   b)
Input	Output
2	8
5	20
7	28

   c)
Input	Output
2	6
4	10
9	20

4. Look at the number line.

   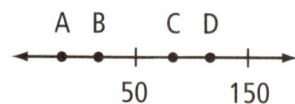

   a) What number could each letter represent? Explain your reasoning.

   b) Place 60 on the number line.

   c) Is C greater than or less than 100? Explain your reasoning.

   d) How far apart are A and B? Explain.

Name: _____ Date: _____

1.1 Perimeters of Two-Dimensional Shapes

Student Text pp. 12–17

Key Ideas Review

Draw a line to match each statement in Column A with the best answer in Column B.

A

1. Total distance around a two-dimensional shape
2. Perimeter of a square
3. Perimeter of a rectangle
4. Perimeter of any two-dimensional shape
5. Measurement units for perimeter

B

a) 4 × side length
b) Centimetres, metres, kilometres
c) Perimeter
d) 2 × length + 2 × width
e) sum of the side measures

Example: Add a Border to a Room

The Lees are adding a paper border around their recreation room.

a) How much border is needed?

b) The border paper costs $2.75/m. What is the total cost of the border?

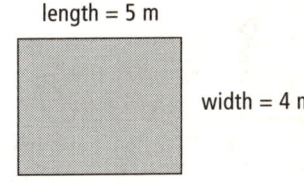

length = 5 m

width = 4 m

Solution

b) **Method 1: Add Side Lengths**

$P = 5 + 4 + 5 + 4$
$P = 18$
The Lees will need 18 m of border.

Method 2: Use a Formula

$P = (2 \times l) + (2 \times w)$
$P = (2 \times 5) + (2 \times 4)$
$P = 10 + 8$
$P = 18$
The Lees will need 18 m of border.

b) Multiply the perimeter of the room by the cost of the border per metre.
Cost = 18 × 2.75
Cost = 49.50
The total cost of the border will be $49.50.

Name: _____   Date: _____

Practise

1. Determine the perimeter of each polygon.

Shape	Side Measures (cm)	Perimeter (cm)
Triangle	2.3, 4.5, 1.7	
Isosceles triangle	Base = 1.5, Other sides = 2.1	
Equilateral triangle	4.7	
Quadrilateral	3.1, 2.6, 1.9, 4.0	
Square	5	
Rectangle	7.8, 4.3	
Parallelogram	2.9, 4.8	
Regular pentagon	3.7	
Isosceles trapezoid	2.1, 0.6, Other sides = 3.8	
Regular hexagon	5.9	

Literacy Connections

Isoceles means two sides are equal.

Equilateral or regular means all sides are equal.

2. Each perimeter in question 1 represents a letter.

Triangle perimeter = K   Isosceles triangle perimeter = A
Equilateral triangle perimeter = L   Quadrilateral perimeter = W
Rectangle perimeter = E   Square perimeter = C
Parallelogram perimeter = B   Regular pentagon perimeter = I
Isosceles trapezoid perimeter = H   Regular hexagon perimeter = T

Each perimeter is shown under a blank. Replace the number with its letter for the answer to this riddle.

What do a zebra and a skunk have in common?

Answer: They both like

___ ___ ___ ___ ___ & ___ ___ ___ ___ ___
15.4 14.1 5.7 20 8.5   11.6 10.3 18.5 35.4 24.2

Study Skills

Look for this feature throughout the book. It will provide ideas to help you prepare for a chapter test.

3. Each day, Raj must run around the school five times as part of the school's fitness program. How far does Raj run each day

   a) in metres?
   b) in kilometres?

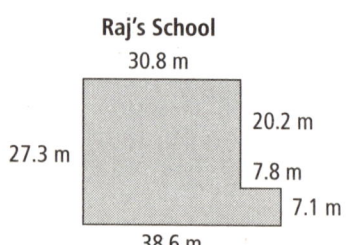
Raj's School

4. How could you find the perimeter of your school? Explain.

2   MHR • Chapter 1: Measurement and Number Sense

Name: _____ Date: _____

1.2 Area of a Parallelogram
Student Text pp. 18–21

Key Ideas Review

1. Label the base and height of each parallelogram.

a) 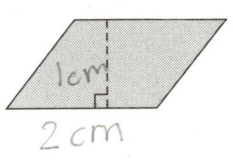 b) 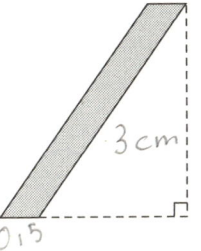 c)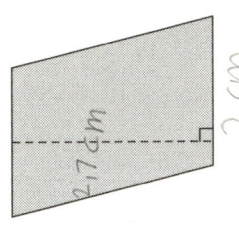

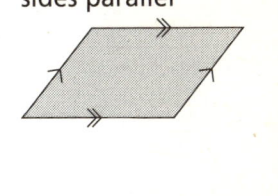

parallelogram
- four-sided figure with both pairs of opposite sides parallel

2. Complete the area formula for a parallelogram.

Area = __b__ × __h__

Example: Parallelogram Area

The parallelogram is drawn on centimetre grid paper.
Calculate the area of the parallelogram.

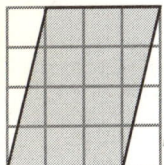 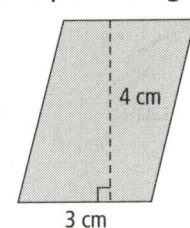

$A = b \times h$
$= 4 \text{ cm} \times 3 \text{ cm}$
$= 12 \text{ cm}^2$

Solution

$A = b \times h$
$A = 3 \times 4$
$A = 12$

The area of the parallelogram is 12 cm².

Practise

1. Each parallelogram is drawn on centimetre grid paper. Determine the areas.

a) 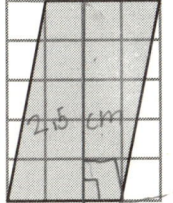 b)  c)

$A = b \times h$
$= 1.5 \text{ cm} \times 2.5 \text{ cm}$
$= 3.75 \text{ cm}^2$

$A = B \times h$
$= 1 \text{ cm} \times 1 \text{ cm}$
$= 1 \text{ cm}^2$

$A = B \times h$
$= 1.5 \text{ cm} \times 0.5 \text{ cm}$
$= 0.75 \text{ cm}^2$

1.3 Area of a Triangle

Student Text pp. 22–25

Key Ideas Review

Draw a line to match each item in Column A with the correct item in Column B.

A

1. Area of a triangle formula
2. Type of units used for measuring area
3. Polygon whose area is half the area of parallelogram
4. Angle that is always formed by the base and height of a triangle

B

- right
- $A = b \times h \div 2$
- triangle
- square

Example: Apply the Triangle Area Formula

The triangle was drawn on metre grid paper.
Find the area of the triangle.

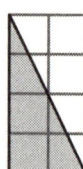

Solution

- base = 2.0 m
- height = 3.5 m

$A = b \times h \div 2$
$A = 2.0 \times 3.5 \div 2$
$A = 7.0 \div 2$
$A = 3.5$

You can also use $A = \dfrac{b \times h}{2}$.

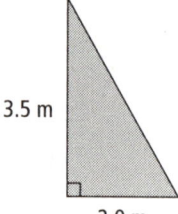

The area of the triangle is 3.5 m².

Practise

1. Identify the base and height of each triangle.

a)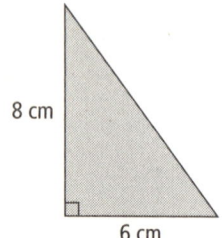

base = __6 cm__
height = __8 cm__

b)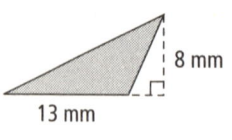

base = __13 mm__
height = __8 mm__

c)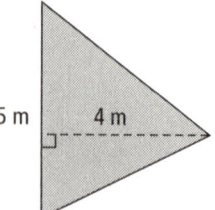

base = __5 m__
height = __4 m__

d)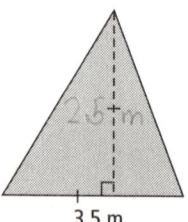

base = __3.5 m__
height = __2.5 m__

Name: _____   Date: _____

2. a) Find the area of each triangle in square centimetres.

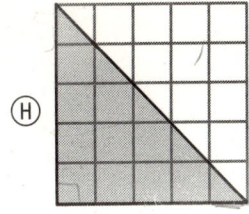
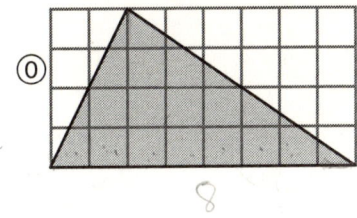
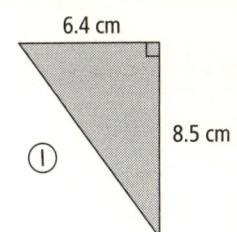

A = 12.5 cm²   A = 16 cm²   A = 27.2 cm²

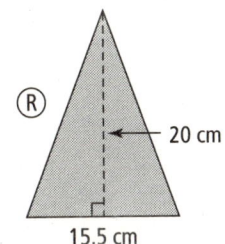
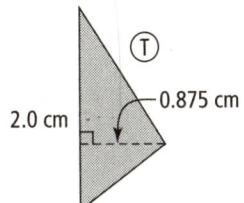
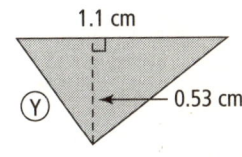

A = 155 cm²   A = 0.875 cm²   A = 0.2915 cm²

b) Each number under a blank is a triangle area. Replace the number with the matching triangle letter shown in the circle.

At what time do dentists prefer to make appointments?

Answer:

A   O   O   A   H   -   A   H   I   R   A   Y
0.875 16 16 0.875 12.5 0.875 12.5 27.2 155 0.875 0.2915

Apply

3. How is the area of a right triangle related to the area of a rectangle? Use diagrams, number, and symbols to explain.

1.4 Apply the Order of Operations

Student Text pp. 26–29

Key Ideas Review

1. Find all of the operations of BODMAS in the word search.

Q	U	S	A	A	D	L	B	M
A	M	U	L	T	I	P	L	Y
B	R	A	C	K	V	R	S	L
O	R	D	E	R	I	O	C	P
D	R	A	O	R	D	E	A	A
A	E	P	C	S	E	D	R	L
S	R	A	C	K	E	T	T	E
A	D	D	E	K	E	R	B	M
C	O	R	D	A	D	T	U	R
T	C	A	R	T	B	U	S	M

2. List the correct sequence of steps for a calculation using the words in question 1.

_____

_____

_____

_____

_____

Example: The Order of Operations

Evaluate. $2 \times 6 - (3.2 - 0.4) \div 0.7 + 3 \times 5$

Solution

Method 1: Pencil and Paper

$2 \times 6 - (3.2 - 0.4) \div 0.7 + 3 \times 5$ Brackets
$= 2 \times 6 - 2.8 \div 0.7 + 3 \times 5$ Multiply, Divide, Multiply
$= 12 - 4 + 15$ Subtract
$= 8 + 15$ Add
$= 23$

Method 2: Calculator

$2 \times 6 - (3.2 - 0.4) \div 0.7 + 3 \times 5$ Brackets: C 3.2 − 0.4 = 2.8
$= 2 \times 6 - 2.8 \div 0.7 + 3 \times 5$ Multiply and Divide: C 2 × 6 = 12, C 2.8 ÷ 0.7 = 4, C 3 × 5 = 15

$= 12 - 4 + 15$ Add and Subtract: C 12 − 4 + 15 = 23
$= 23$

Study Skills

Throughout this workbook, you will see several ways to solve the same problem. Use one way to solve the problem. Use the other way to check.

Name: _____   Date: _____

Practise

1. a) Evaluate each expression.
   b) Find a path from Start to Finish, moving horizontally or vertically but not diagonally. Each new box must have a value greater than the previous box.
   c) Check your answers using a calculator.

START

$99 \div 11 \div 3$	$6 \div 3 + 2 \times 4$	$(4 + 3) \times 2 - 2$	$12 \div (14 \div 7) + 6$
$9 + (13 - 10) \div 3$	$4 + 3 \times 2$	$15 \div 5 - 1$	$10 \div 2 + 3 \times 7$
$28 \div (13 - 11)$	$(2.1 - 0.9) \times 20$	$(10 - 7) \div 3 + (9 - 3) \times 4$	$20 \times (0.9 \div 3) - 2 \times 1 + (9 - 8)$
$8 \times 9 \div 4$	$10 \times (0.9 \div 0.3) - 7 \times 1 + (9 - 6)$	$(14.6 + 1.4) \times (2.9 - 0.4)$	$(9 + 2) \times 4$

FINISH

Apply

2. What's wrong? To claim a prize, Brad answers a skill-testing question.

   $100 \times 20 \div 5 - 4 + 8 \times 3$ $100 \times 20 \div 5 - 4 + 8 \times 3$
   $= 100 \times 4 - 4 + 8 \times 3$ $=$
   $= 100 \times 0 + 24$ $=$
   $= 0 + 24$ $=$
   $= 24$ $=$

   a) Circle the error in Brad's solution.
   b) Give a correct solution. Show your work in the space provided.

1.5 / 1.6 Area of a Trapezoid / Draw Trapezoids

Student Text pp. 30–39

Key Ideas Review

1. **a)** Label the base, the side opposite the base, and the height of the trapezoid. Use the following letters: *b* for the base, *a* for the opposite side, and *h* for the height.

   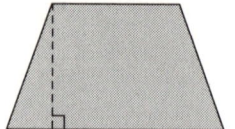

   trapezoid
   - four-sided figure with one pair of opposite sides **parallel**

   b) Show which sides are parallel by adding arrowheads (>).

2. Complete the formulas for the area of a trapezoid.

   Area of trapezoid = (side *a* + _____) × _____ ÷ 2   (Use words.)

   $A = (a + \_\_\_) \times \_\_\_ \div 2$ (Use letters.)

3. You can also find the area of a trapezoid by dividing it into two _____ and finding the area of each.

Example 1: Trapezoid Area

The trapezoid was drawn on centimetre grid paper. Find the area of the trapezoid.

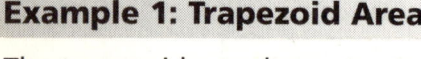

Solution

Method 1: Split Into Triangles

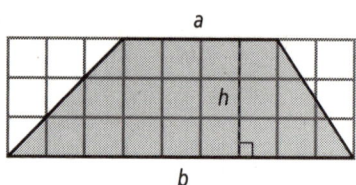

Area of Triangle 1 = 3 × 4 ÷ 2
            = 6

Area of Triangle 2 = 9 × 3 ÷ 2
            = 13.5

Area of trapezoid = Area of Triangle 1 + Area of Triangle 2
            = 6 + 13.5
            = 19.5

The area of the trapezoid is 19.5 cm².

8   MHR • Chapter 1: Measurement and Number Sense

Method 2: Use a Formula

$A = (a + b) \times h \div 2$

$= (4 + 9) \times 3 \div 2$

$= 13 \times 3 \div 2$

$= 39 \div 2$

$= 19.5$

You could also use $A = \dfrac{(a + b) \times h}{2}$.

Brackets: C 4 + 9 = 13
Multiply: C 13 × 3 = 39
Divide: C 39 ÷ 2 = 19.5

The area is 19.5 cm².

Method 3: Estimate

There are about 19 centimetre squares in the trapezoid, so 19.5 cm² is a reasonable answer.

Example 2: Draw a Trapezoid Given Its Area

Draw a trapezoid with an area of 30 cm².

Solution

- Use centimetre grid paper to draw a rectangle with an area of 30 cm².

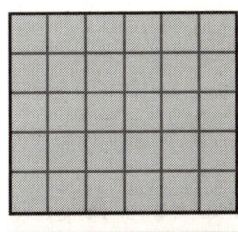

- Split the rectangle into a triangle and trapezoid.

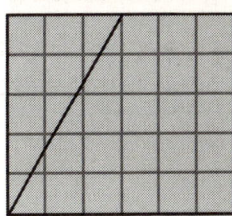

- Rearrange the shapes to make a trapezoid. Measure the sides of the trapezoid. Add the measurements to the diagram.

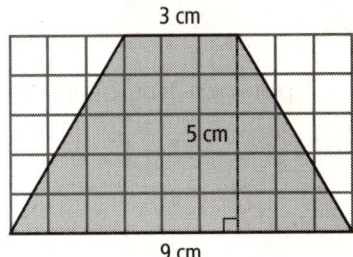

The area of the trapezoid should be about 30 cm².

Check by using the area formula for a trapezoid.

Practise

1. The trapezoid is drawn on centimetre grid paper.

   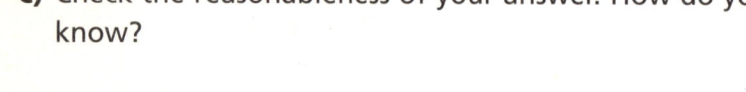

   a) Find the perimeter.

   b) Find the area.

   c) Check the reasonableness of your answer. How do you know?

2. a) Draw a trapezoid with a perimeter of 20 cm. Label the sides.

   b) What is the area of your trapezoid?

   Area =

   c) Use another method to check your answer. Explain or show your work.

3. a) Find the perimeter of the pentagon.

   Perimeter =

   b) Calculate the area of the pentagon.

   Area =

   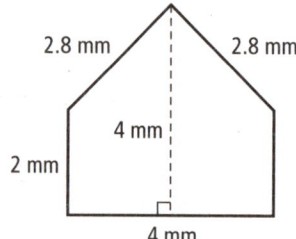

   Hint

   The pentagon is made up of two congruent trapezoids. The height of one trapezoid is half the base of the pentagon.

Apply

4. Ribbon is used to trim the flag.

   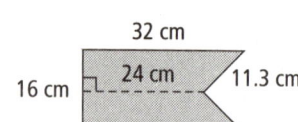

   a) How much material is needed to make one flag? Round your answer to the nearest square centimetre.

   Amount of material ≑

   b) How much ribbon is needed to trim one flag?

   Length of ribbon =

10 MHR • Chapter 1: Measurement and Number Sense

1.7 Composite Shapes

Student Text pp. 40–45

Key Ideas Review

Fill in the blanks and label the diagram. Use the words from the list. Two words may be used more than once.

| distance | triangle | area | trapezoid |
| perimeter | rectangle | composite | simpler |

1. A _____ shape is a two-dimensional shape.

   It is made up of two or more _____ shapes.

2. Label the different shapes in the pentagon.

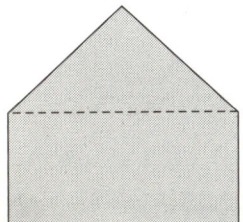 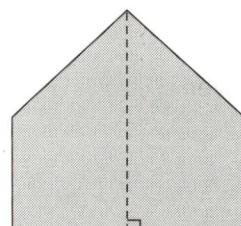

3. Describe two ways to find the area of the pentagon.

4. The _____ of the pentagon is the total _____ around it.

Example: Perimeter and Area of a Garden

a) Find the perimeter of the garden.

b) Find the area of the garden.

Solution

a) Perimeter = 12.2 + 5 + 3.3 + 5.6 + 5.6 + 5.6 + 3.3 + 5
   = 45.6
   The perimeter is 45.6 m.

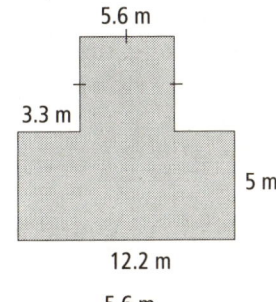

b) The garden is made up of a square and a rectangle.
   Area = Area of rectangle + Area of square
   = (5 × 12.2) + (5.6 × 5.6)
   = 92.36
   The area is 92.36 m².

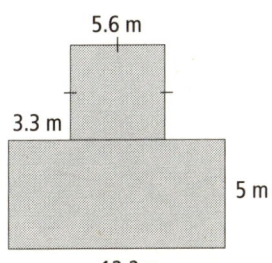

Practise

1. Complete the chart.

Composite Shape	Simpler Shapes	Perimeter (cm)	Area (cm²)
(triangle composite with measurements 3.5, 3, 4, 2.5, 2, 8)			
(fish-shaped composite with measurements 6, 3, 2, 10, 3, 2, 6)			

Apply

2. Which backyard would cost more

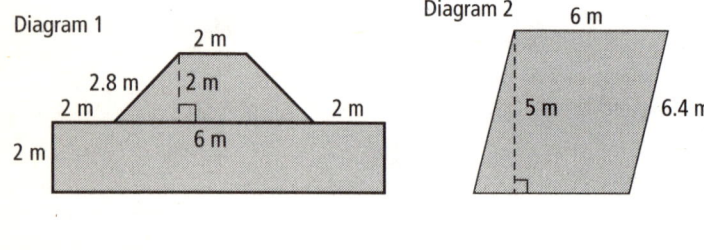

   a) to cover in sod? Diagram _____

   b) to fence in? Diagram _____

   Give reasons.

Making Connections

1. Make a scale drawing of your bedroom floor.
2. Measure the floor and add these measurements to your diagram.
3. Calculate the perimeter and area of the floor.
4. Research newspaper advertisements or go to your local home repair store to select a covering for your floor.

How much would it cost to cover the floor?

Name: _____   Date: _____

Chapter 1: Reviewing for the Test

1.1 Perimeters of Two-Dimensional Shapes
page 10

1. For each shape, determine the perimeter.

   a)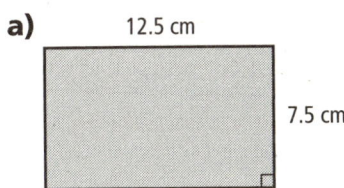

   Perimeter =

   b)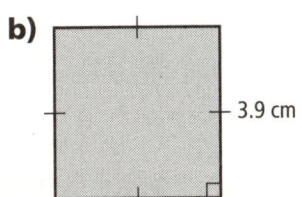

   Perimeter =

   c)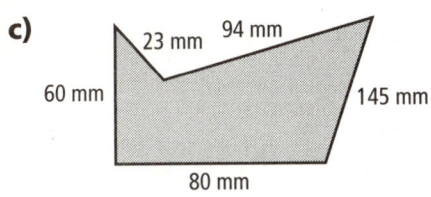

   Perimeter =

2. Fencing costs $8.50/m. What is the cost of fencing for this playground?

   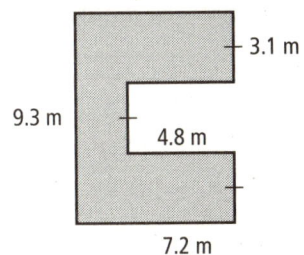

   A $207.40   B $260.10
   C $362.10   D $192.95

1.2 Area of a Parallelogram
page 18

3. a) Draw a parallelogram with a base of 4 cm and a height of 3 cm on the centimetre grid paper. Label the diagram with the measures.

   b) Calculate the area of the parallelogram.

   Area =

1.3 Area of a Triangle
page 22

4. Calculate the area of each triangle in square centimetres.

   a)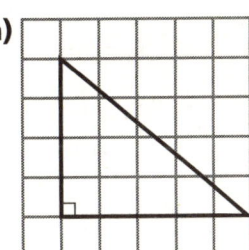

   Area =

   b)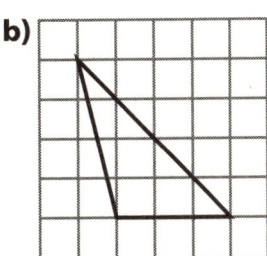

   Area =

Name: _____ Date: _____

1.4 Apply the Order of Operations
page 26

5. Evaluate each expression.

   a) 4 + 3 × 5 + (7 − 3) =

   b) 9 ÷ (2 + 1) − 2 =

   c) 10 ÷ (2 + 3) × 4 =

   d) 7 × 4 + (3 − 2) + 6 =

   e) 18 − (6 × 2) + 7 ÷ (10 − 3) =

1.5 Area of a Trapezoid
1.6 Draw Trapezoids
page 30

6. The trapezoid is drawn on centimetre grid paper. Calculate the area of the trapezoid.

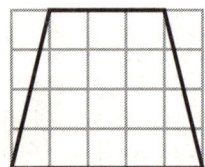

   Area =

7. Draw a trapezoid with a base of 4 mm, height of 3 mm, and area of 15 mm² on the millimetre grid paper. Label the diagram with the measures.

1.7 Composite Shapes
page 40

8. The Roys want to plant a single row of pansies along the edge of the entire backyard, with one plant for every 1 m.

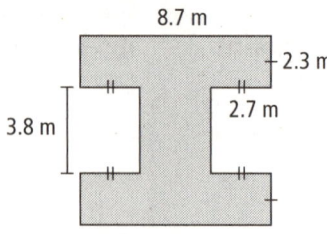

   a) How many plants will they need?

      Number of plants =

   b) Each plant costs 59¢. How much will all of the needed plants cost?

      Cost =

   c) The Roys may replace the sod in their entire backyard. How much sod will they need?

      Amount of sod =

9. Jeremy created this birthday card for his grandfather on centimetre grid paper.

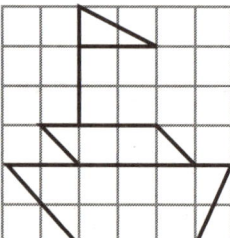

   All parts of the boat are made of paper except for the flagpole.

   How much paper did Jeremy use to create the boat?

   Amount of paper =

14   MHR • Chapter 1: Measurement and Number Sense

Name: _____ Date: _____

2.1 Classify Triangles

Student Text pp. 54–59

Key Ideas Review

Complete the crossword puzzle using the words from the list.

ray	transformations	scalene	vertex	turn
line segment	flip	protractor	equilateral	slide
triangle	isosceles	obtuse	acute	right

Across

2. Name of triangle with no equal sides.
7. Name of triangle that has all equal sides.
9. Name of triangle with all three angles less than 90°.
11. Another name for a rotation.
12. Another name for a translation.
13. Where two rays meet.
15. What slides, flips, and turns are.

Down

1. Name of triangle that has two equal sides.
3. A three-sided shape.
4. Name of triangle with angle greater than 90°.
5. Name of triangle that has 90° angle.
6. Instrument for measuring the size of an angle.
8. Part of a line that connects two points.
10. Another name for a reflection.
14. Name of one arm of an angle.

Literacy Connections

Did you know that the first three, four, or five letters in the name of a specific polygon tell you how many sides the polygon has?

tri = 3, quad = 4, penta = 5, hexa = 6

Use your dictionary to find other prefixes and their meanings.

Name: _____ Date: _____

Practise

1. Classify each triangle as equilateral, isosceles, or scalene. Explain your choice.

   a)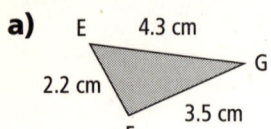

   b)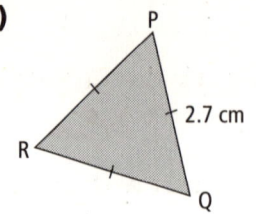

   c)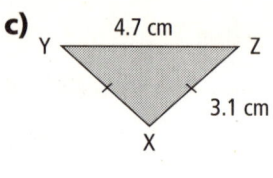

   _____   _____   _____

   _____   _____   _____

   _____   _____   _____

2. Classify each triangle as acute, right, or obtuse. Explain your choice.

   a)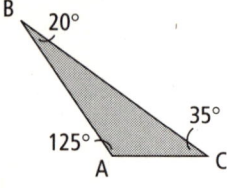

   b)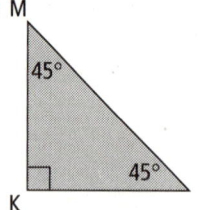

   c)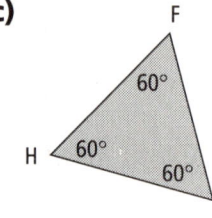

   _____   _____   _____

   _____   _____   _____

   _____   _____   _____

3. Classify each triangle in question 2 by its side lengths. Explain your choice.

   a) _____   b) _____   c) _____

   _____   _____   _____

   _____   _____   _____

4. Name all of the different triangles in this drawing.

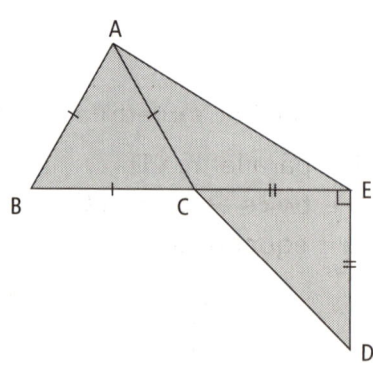

Equilateral	Isosceles	Scalene

Acute	Right	Obtuse

16   MHR • Chapter 2: Two-Dimensional Geometry

Name: _____ Date: _____

2.2 Classify Quadrilaterals
Student Text pp. 60–65

Key Ideas Review

Draw a line to match each word in Column A with its description in Column B.

A
1. quadrilateral
2. kite
3. parallelogram
4. square
5. rhombus
6. trapezoid

B
a) quadrilateral with all sides equal, all angles 90°
b) quadrilateral with all sides equal
c) quadrilateral with opposite pairs of sides parallel
d) quadrilateral with two pairs of equal sides, none parallel
e) quadrilateral with one pair of parallel sides
f) four-sided polygon

Practise

1. Classify each quadrilateral.

   a)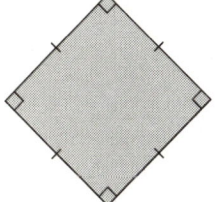

   b)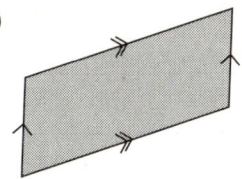

   c)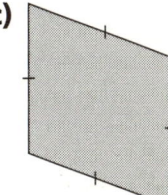

   d)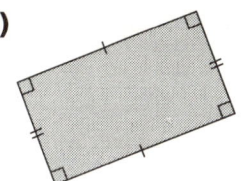

   e)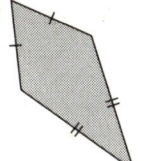

   f)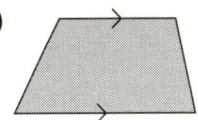

2. Draw and label a quadrilateral that matches each description, then classify it.

   a) AB is parallel to CD.
   AB is twice as long as CD.
   AD is equal to BC.

   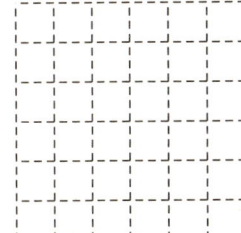

   b) EF is parallel and equal to HG.
   EH is parallel and equal to FG.
   ∠E = 90°, ∠F = 90°.

2.3 / 2.4 Congruent Figures / Congruent and Similar Figures

Student Text pp. 66–75

Key Ideas Review

Fill in the blanks with the words from the list. Each numbered grey box corresponds to a letter that spells the answer to the riddle below.

> sides   congruent   similar   lengths
> same   angles   proportion

1. Two figures that have the same shape and the same size are __ __ __ __ __ __ __ __ __ .
   (1)

2. Corresponding parts of congruent figures have the __ __ __ __ measures.
   (2)

3. Two figures that have the same shape but may be different sizes are __ __ __ __ __ __ __ .
   (3)

4. Corresponding __ __ __ __ __ __ in similar figures are equal.

5. The __ __ __ __ __ __ __ of corresponding __ __ __ __ __ __
   (4)   (5)

   in similar figures are in __ __ __ __ __ __ __ __ __ __ .
                              (6)   (7)

Which is faster, hot or cold?

Answer:

___ ___ ___ because you can ___ ___ ___ ___ ___ ___ ___ ___ ___ ___ ___ ___
 4   6   7   1   2   7   1   4   2   1   6   3   5

Name: _____   Date: _____

Example 1: Matching Parts of Congruent Figures

Compare quadrilateral ABCD and quadrilateral FGHI.

a) List the corresponding angles and sides.
b) Are the two quadrilaterals congruent? Explain your answer.

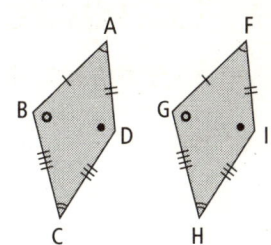

Solution

a)   Corresponding angles   Corresponding sides

         ∠A = ∠F   AB = FG
         ∠B = ∠G   BC = GH
         ∠C = ∠H   CD = HI
         ∠D = ∠I   AD = FI

b) The corresponding angles and side lengths are equal. Therefore, quadrilateral ABCD and quadrilateral FGHI are congruent.

Example 2: Matching Parts of Similar Figures

Compare the corresponding angles and the corresponding sides of quadrilateral ABCD with quadrilateral FGHI.

Are the two quadrilaterals similar?

Explain your answer.

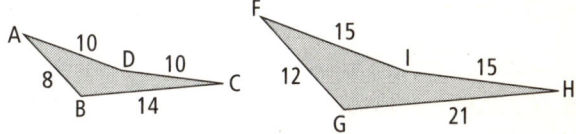

Solution

Corresponding angles   Corresponding sides

    ∠A = ∠F $\dfrac{AB}{FG} = \dfrac{8}{12}$

    ∠B = ∠G $\dfrac{BC}{GH} = \dfrac{14}{21}$

    ∠C = ∠H $\dfrac{CD}{HI} = \dfrac{10}{15}$

    ∠D = ∠I $\dfrac{AD}{FI} = \dfrac{10}{15}$

Quadrilateral ABCD and quadrilateral FGHI are similar. They have the same shape, the corresponding angles are equal, and their sides are in proportion. Each side of quadrilateral FGHI is one and a half times the corresponding side of quadrilateral ABCD.

Practise

1. Compare the triangles. Explain your answers.

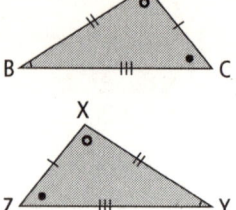

a) Which angles correspond between the triangles?
∠A = ∠   ∠B = ∠   ∠C = ∠

b) Which sides correspond between the triangles?
AB =   BC =   AC =

c) △ABC and △XYZ are _____.

2. Compare the triangles. Explain your answers.

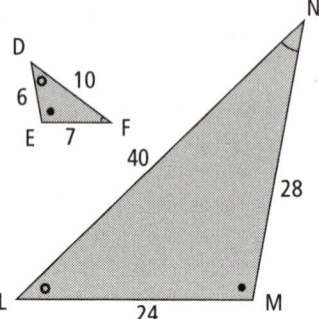

a) Which angles correspond between the triangles?
∠D = ∠   ∠E = ∠   ∠F = ∠

b) How are corresponding sides related?

$$\frac{DE}{LM} = \frac{EF}{MN} = \frac{DF}{LN} =$$

c) △DEF and △LMN are _____.

Apply

3. If two triangles have the same shape and area, must they be congruent? Draw diagrams to illustrate your answer.

4. Is this statement true? "All equilateral triangles are similar." Explain your decision. Draw diagrams to illustrate your answer.

Chapter 2: Reviewing for the Test

2.1 Classify Triangles
page 54

1. Classify each triangle in two ways.

a)

b)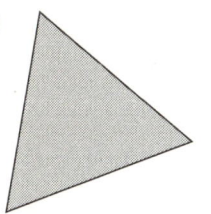

c)

2.2 Classify Quadrilaterals
page 60

2. Classify each quadrilateral.

a) 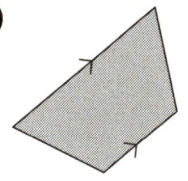   b)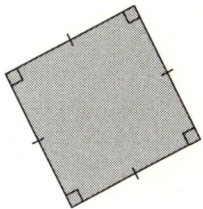

c) 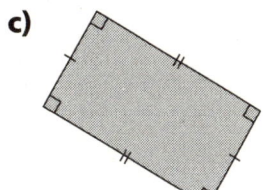   d)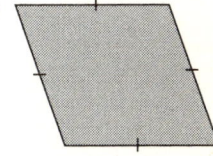

e) 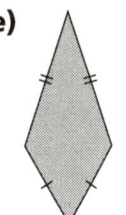   f)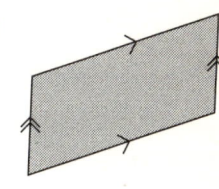

Chapter 2 Reviewing for the Test • MHR   21

2.3 Congruent Figures
2.4 Congruent and Similar Figures

page 66

3. Complete the charts. Are the figures congruent or similar?

a)

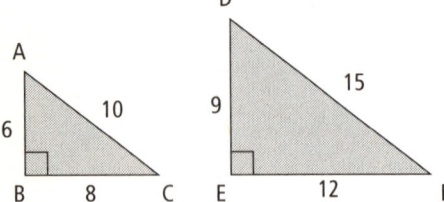

Corresponding

Angles   Sides

∠A = ∠ $\dfrac{AB}{DE} =$

∠B = ∠ $\dfrac{BC}{EF} =$

∠C = ∠ $\dfrac{AC}{DF} =$

△ABC and △DEF are _____.

b)

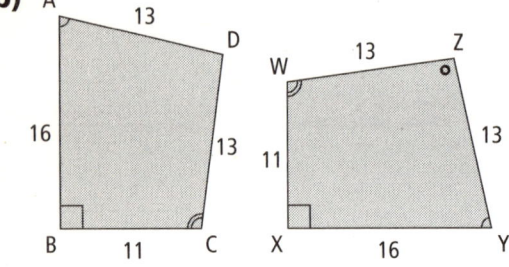

Corresponding

Angles   Sides

∠A = ∠   AB =

∠B = ∠   BC =

∠C = ∠   CD =

∠D = ∠   AD =

ABCD and WXYZ are _____.

4. Draw two figures that are congruent. Add measurements to your diagrams. (Do not draw triangles or quadrilaterals.) Explain why they are congruent.

5. Draw two figures that are similar. Add measurements to your diagrams. (Do not draw triangles or quadrilaterals.) Explain why they are congruent.

Name: _____   Date: _____

3.1 Add Fractions Using Manipulatives

Student Text pp. 86–89

Key Ideas Review

1. **a)** Draw lines on Shape A to show how many of Shape B will fit.

   b) What fraction of Shape A does Shape B represent? Write the fraction on Shape B. Use pattern blocks to help you if needed.

Shape A (Represents 1)	Shape B
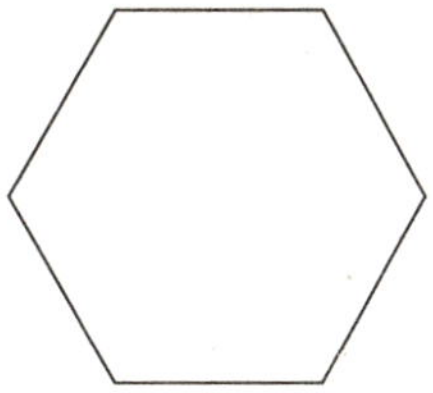 hexagon	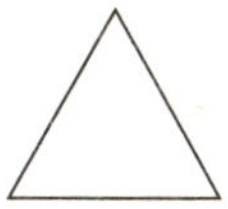 triangle
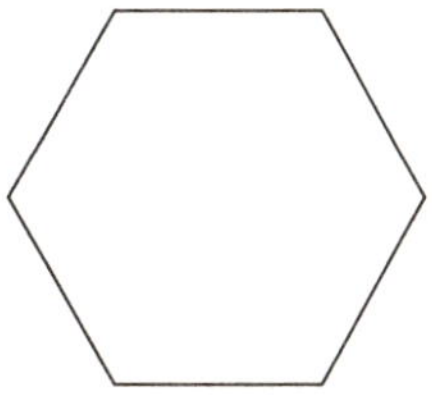 hexagon	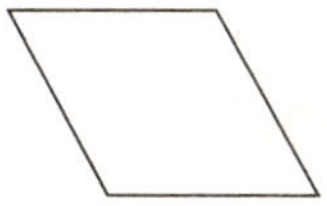 rhombus
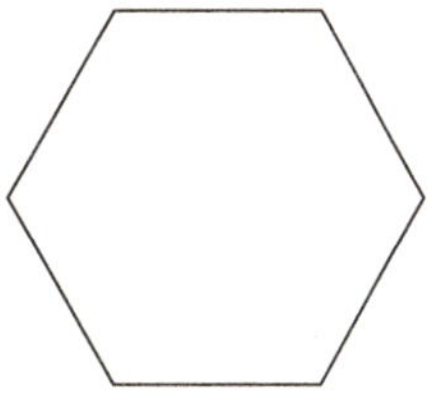 hexagon	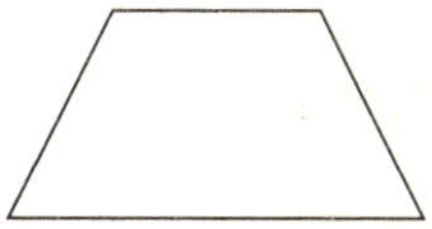 trapezoid
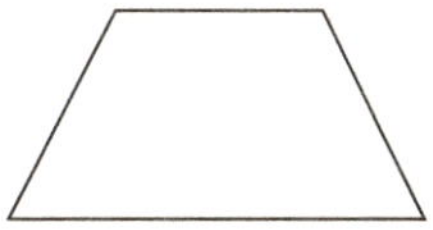 trapezoid	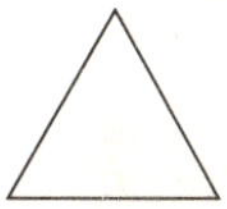 triangle
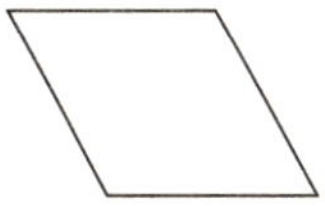 rhombus	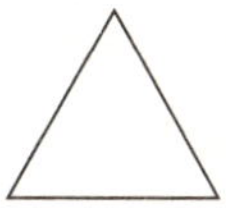 triangle

Making Connections

Create your own pattern blocks.

Trace at least four regular hexagons onto paper.

Colour one hexagon yellow. This one will represent a whole.

Colour one hexagon green. Divide this one into six congruent triangles.

Colour another hexagon blue. Divide this one into three congruent rhombuses.

Colour the last hexagon red. Divide this one into two congruent trapezoids.

Cut out the separate shapes in each hexagon.

Name: _____ Date: _____

Example: Add Fractions

Add $\frac{1}{6} + \frac{2}{3}$.

Solution

Method 1: Use Concrete Materials

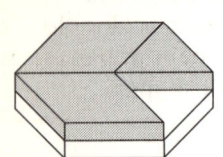 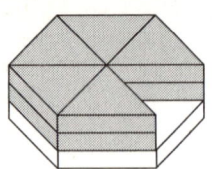

Method 2: Use a Diagram

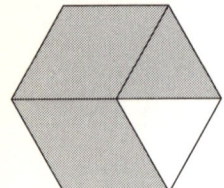 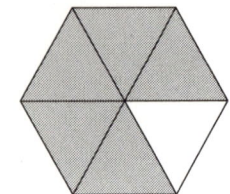

equivalent fractions
- represent the same part of the whole or group.
- In the example, the two rhombuses cover $\frac{2}{3}$ of the hexagon.
- The two rhombuses can be covered by four triangles. The four triangles represent $\frac{4}{6}$.
- Since the two rhombuses cover the same area as four triangles, $\frac{2}{3}$ and $\frac{4}{6}$ are equivalent fractions.

Practise

1. Shade the hexagon to represent the fraction.

   a) $= \frac{1}{6}$ b) $= \frac{1}{3}$ c) $= \frac{1}{2}$

   d) $= \frac{2}{3}$ e) $= \frac{5}{6}$ f) $= \frac{6}{6}$

2. Add. Show the result using a diagram.

   a) $\frac{1}{6} + \frac{3}{6}$ 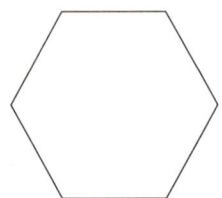   b) $\frac{1}{6} + \frac{1}{3}$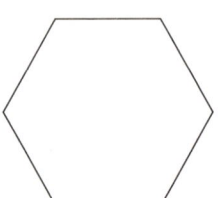

24 MHR • Chapter 3: Fraction Operations

Name: _____ Date: _____

3. Write an addition sentence to describe how many hexagons are covered in each of the following.

Hint
The denominator in each fraction should show the total number of congruent shapes in the hexagon.
The numerator in each fraction should show the number of shaded shapes.

a)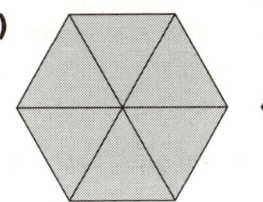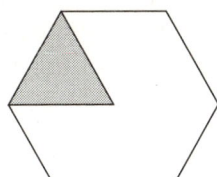

b)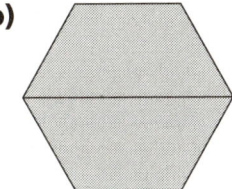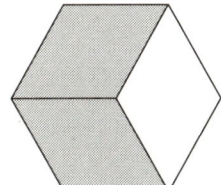

c)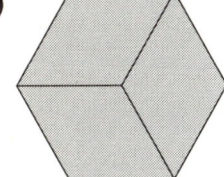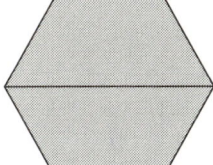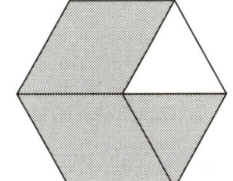

Apply

4. Tam had three quarters, and was given two more quarters.

   a) Write an addition sentence to show the amounts.

   b) Express the total amount of money he has as a mixed number.

   c) Write the mixed number as a dollar amount.

Study Skills

Create different concrete materials, such as fraction strips (using centimetre grid paper) or fraction circles. Use one type of concrete material to model the addition sentences. Use another type of concrete material to check your answers.

Name: _____ Date: _____

3.2 Subtract Fractions Using Manipulatives

Student Text pp. 90–93

Key Ideas Review

1. Order from 1 to 4 the steps for subtracting fractions.

   Use the diagrams for each step to help you.

   ___ Identify the fraction that remains.

   $\frac{2}{3} - \frac{1}{6} = \frac{4}{6} - \frac{1}{6}$
   $= \frac{3}{6}$

   ___ Represent each fraction using parts of equal size.

   $\frac{2}{3} = \frac{4}{6}$ $\frac{1}{6}$

   ___ Remove the parts represented by the fraction being subtracted.

   $\frac{2}{3} - \frac{1}{6} = \frac{4}{6} - \frac{1}{6}$

   ___ Express the difference in lowest terms.

   $\frac{3}{6} = \frac{1}{2}$

Example: Subtract Fractions

Subtract $\frac{5}{6} - \frac{1}{2}$.

Solution

Method 1: Use Manipulatives

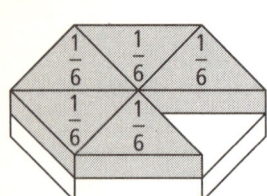 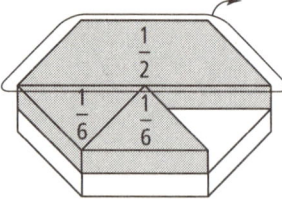 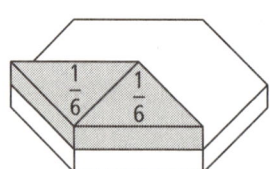 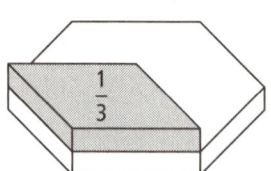

Method 2: Use a Diagram

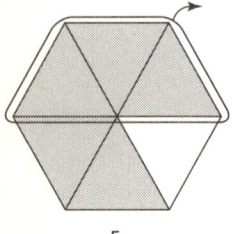 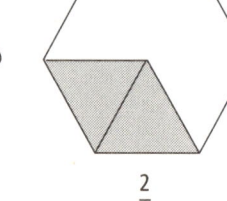 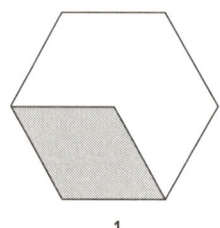

$\frac{5}{6}$ $\frac{2}{6}$ $\frac{1}{3}$

26 MHR • Chapter 3: Fraction Operations

Name: _____   Date: _____

Practise

1. Write a subtraction statement to represent each diagram. Then draw the result.

a) 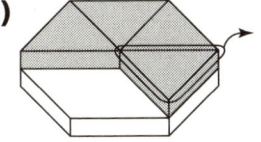 _____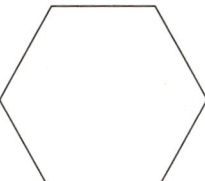

b) 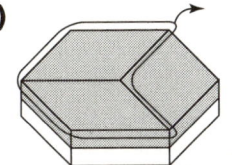 _____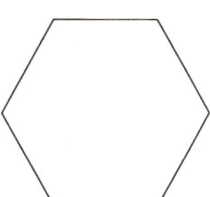

c) 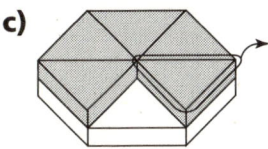 _____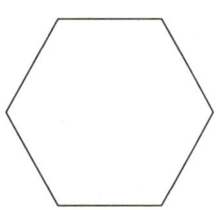

2. Use diagrams to show each subtraction. Use pattern blocks to check. Then write the result.

a) $1 - \dfrac{1}{6} =$ _____

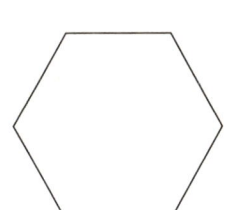 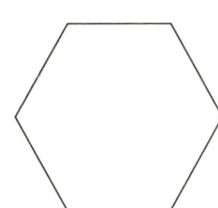

b) $\dfrac{5}{6} - \dfrac{3}{6} =$ _____

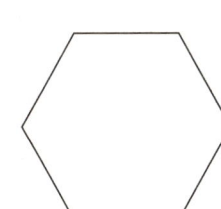 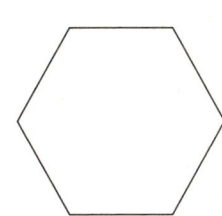

c) $\dfrac{2}{3} - \dfrac{1}{6} =$ _____

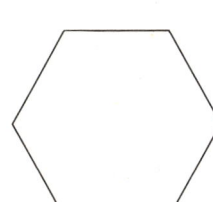 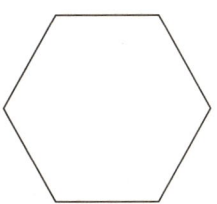

d) $1 - \dfrac{2}{3} =$ _____

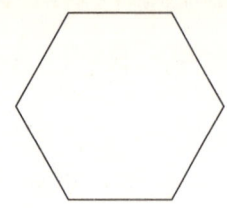
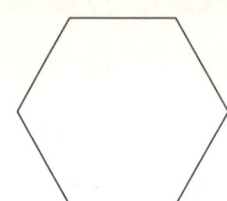

Apply

3. Suppose 1 large square = 1 whole. Write a subtraction statement to represent this diagram.

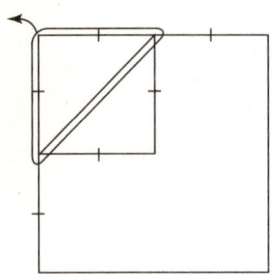

4. Janine has one dozen eggs. She used two eggs to make cookies. She used three eggs to make a cake. She had one egg for breakfast.

Literacy Connections
1 dozen = 12
1 baker's dozen = 13

a) How many eggs are left?

b) What fraction does the remaining number of eggs represent?

5. Mrs. Mason bought two baker's dozen of cookies for her staff. Her staff ate 14 cookies. She ate two cookies. She gave eight cookies away.

a) How many cookies are left?

b) What fraction does the remaining number of cookies represent?

Name: _____ Date: _____

3.3 Find Common Denominators

Student Text pp. 94–97

Key Ideas Review

To find a common denominator, you can use paper folding, diagrams, and common multiples.

Find the common denominator of $\frac{3}{5}$ and $\frac{2}{3}$.

1. **Method 1: Use Paper Folding**   **Method 2: Use Diagrams**

   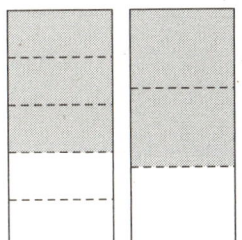   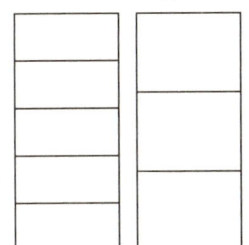

   Add dotted lines to show paper folding.

   Add solid lines to show the parts.

   Write the equivalent fractions.   Write the equivalent fractions.

   $\frac{3}{5} = $ _____ $\frac{2}{3} = $ _____ $\frac{3}{5} = $ _____ $\frac{2}{3} = $ _____

   Hint

   To show $\frac{3}{5}$, fold a piece of paper into fifths one way. Shade three parts. Now fold the paper into thirds the other way.

   To show $\frac{2}{3}$, fold a piece of paper into thirds one way. Shade two parts. Now fold the paper into fifths the other way.

2. **Method 3: Use Multiples**

   a) Circle the common multiples of the denominators.

   b) Put a box around the lowest common multiple of 5 and 3.

   The denominator of $\frac{3}{5}$ is 5.

   Multiples of 5: 5, 10, 15, 20, 25, 30, 35, …

   The denominator of $\frac{2}{3}$ is 3.

   Multiples of 3: 3, 6, 9, 12, 15, 18, 21, 24, 27, 30, …

   multiple
   - the product of a given number and a natural number
   - multiples of 2 are 2, 4, 6, 8, …

Practise

1. List the next five multiples for each pair of numbers.
   Circle the common multiples.
   Put a box around the lowest common multiple.

   a) multiples of 2: 2,

   multiples of 3: 3,

   b) multiples of 3: 3,

   multiples of 4: 4,

Name: _____   Date: _____

2. Complete the diagrams to find a common denominator for each pair of fractions.

   a) $\frac{1}{2}$ and $\frac{1}{6}$

   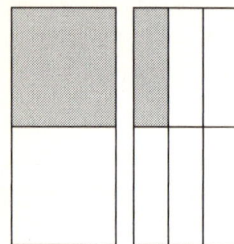

   common denominator: _____

   b) $\frac{1}{2}$ and $\frac{1}{7}$

   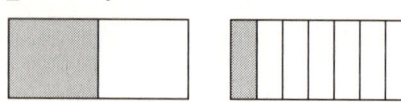

   common denominator: _____

   c) $\frac{3}{4}$ and $\frac{5}{6}$

   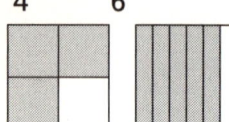

   common denominator: _____

   d) $\frac{2}{8}$ and $\frac{5}{16}$

   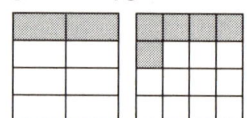

   common denominator: _____

3. For each pair of fractions in question 2, write equivalent fractions. Use a common denominator.

   a) $\frac{1}{6} =$ $\frac{1}{2} =$

   b) $\frac{1}{2} =$ $\frac{1}{7} =$

   c) $\frac{3}{4} =$ $\frac{5}{6} =$

   d) $\frac{2}{8} =$ $\frac{5}{16} =$

Apply

4. In each pair of fractions, which fraction is greater? Use the < or > symbol to complete the sentence. Include diagrams to help you explain.

   a) $\frac{1}{6}$ ◯ $\frac{4}{6}$

   b) $\frac{2}{3}$ ◯ $\frac{5}{6}$

   c) $\frac{3}{4}$ ◯ $\frac{7}{8}$

Study Skills

Write a strategy for adding or subtracting fractions in your own words. Include words like *multiples, common denominator, numerator,* and *reduce.* Have a study partner read the strategy to see if it is clear.

30   MHR • Chapter 3: Fraction Operations

Name: _____ Date: _____

3.4 Add and Subtract Fractions Using a Common Denominator

Student Text pp. 98–103

Key Ideas Review

1. Order from 1 to 4 the steps to add or subtract fractions.

    ____ Add or subtract the numerators.

    ____ Write the equivalent fractions using a common denominator.

    ____ Find a common denominator for the fractions.

    ____ Express the result in lowest terms.

2. Circle the correct word to complete the sentence.
   Repeated addition of the same fraction can be shown as

    a) a single addition.   **b)** a single subtraction.
    c) multiplication.   **d)** division.

Example 1: Add Fractions

How much pie is left?

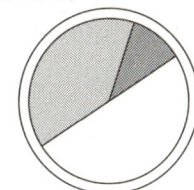

Solution

Method 1: Use Manipulatives

To add fractions, the pieces have to be the same size.

Each triangle represents $\frac{1}{6}$.

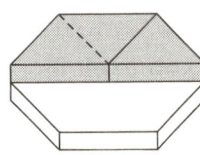

Count the number of pieces.

$\frac{1}{3} + \frac{1}{6} = \frac{3}{6}$ $\frac{3}{6}$ can also be written as $\frac{1}{2}$.

$\frac{3}{6}$ or $\frac{1}{2}$ of a pie is left.

Method 2: Use a Common Denominator

Fold a piece of paper into thirds one way, then in half the other way.

Colour $\frac{1}{3}$ of the page. Two sections are coloured.

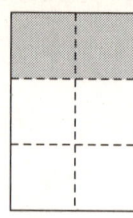

Fold another piece of paper in sixths.

Colour $\frac{1}{6}$ of the page. One section is coloured.

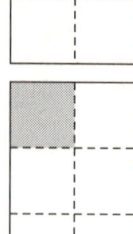

A common denominator is 6.

$\frac{1}{3} + \frac{1}{6} = \frac{2}{6} + \frac{1}{6}$ Add the numerators.

$= \frac{3}{6}$ $\frac{3}{6}$ can also be written as $\frac{1}{2}$.

$\frac{3}{6}$ or $\frac{1}{2}$ of a pie is left.

Example 2: Subtract Fractions

Subtract $\frac{5}{6} - \frac{2}{3}$.

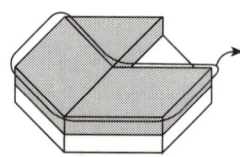

Method 1: Use Manipulatives

To subtract fractions, the pieces have to be the same size.

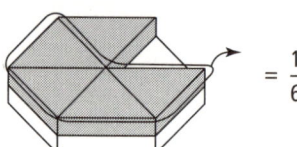

$= \frac{1}{6}$

Method 2: Use Multiples

Multiples of 6: ⑥, ⑫, 18, ...

Multiples of 3: 3, ⑥, 9, ⑫, ...

The first common denominator in the lists is 6.

Write equivalent fractions with 6 as a denominator.

$\frac{5}{6} = \frac{5}{6}$

$\frac{2}{3} = \frac{2 \times 2}{3 \times 2}$ To find equivalent fractions, multiply the numerator and denominator by the same number.

$= \frac{4}{6}$

$\frac{5}{6} - \frac{2}{3} = \frac{5}{6} - \frac{4}{6}$ Subtract the numerators.

$= \frac{1}{6}$

Example 3: Add the Same Fraction

How much pie is left?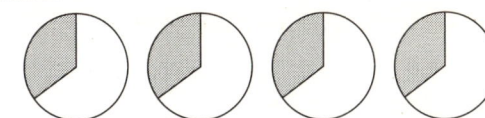

Solution

Method 1: Use Manipulatives

Each rhombus represents $\frac{1}{3}$.

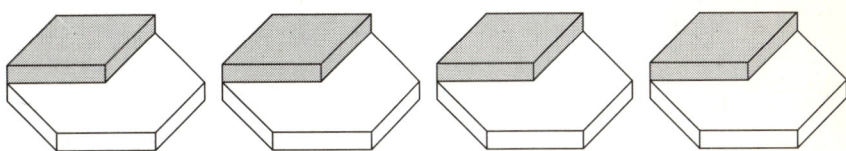

There are four rhombuses.

$\frac{1}{3} + \frac{1}{3} + \frac{1}{3} + \frac{1}{3} = \frac{4}{3}$

Method 2: Multiply

Each strip shows $\frac{1}{3}$.

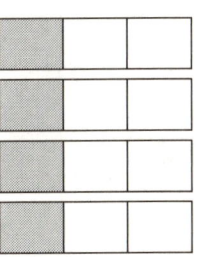

There are four third-strips.

4 thirds = $4 \times \frac{1}{3}$ Multiply the whole number 4 by the numerator.

$= \frac{4}{3}$ $\frac{4}{3}$ can also be written as $1\frac{1}{3}$.

There are $\frac{4}{3}$ or $1\frac{1}{3}$ pies left.

Practise

1. What fraction of each figure is shaded?

   a)    b) 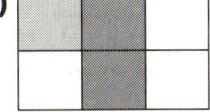   c)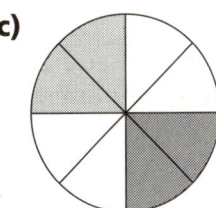

3.4 Add and Subtract Fractions Using a Common Denominator

2. Write an addition sentence for each diagram in question 1.

    a)   b)   c)

3. What fraction of each figure remains?

    a) 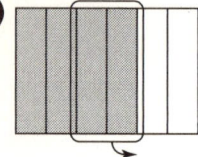   b) 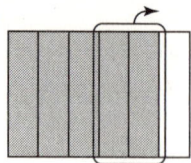   c)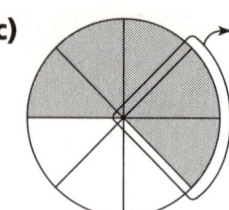

4. Write a subtraction sentence for each diagram in question 3.

    a)   b)   c)

5. In each diagram, what fraction of a whole is shaded?

    a)    b)    c)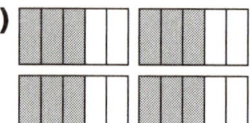

6. Write an addition sentence for each diagram in question 5.

    a)   b)   c)

7. Write a multiplication sentence for each addition sentence in question 6.

    a)   b)   c)

Apply

8. Anna ate $\frac{1}{8}$ of a cake for dessert. John ate $\frac{1}{4}$ and Tom ate $\frac{1}{6}$ of the same cake.

    a) What fraction of the cake was eaten? Use a diagram to help you explain.

    b) Write a subtraction sentence to represent what was left.

3.5 More Fraction Problems

Student Text pp. 104–107

Key Ideas Review

Fill in the blanks with the words from the list.

> check strategy asked own
> diagram different

1. To solve a problem, read what is being _____. **Understand**

   Write it in your _____ words.

2. Decide what _____ to use. **Plan**

   Use a _____ or concrete materials if needed.

3. _____ your answer. **Do It!**

4. Use a _____ strategy to solve the problem. **Look Back**

Example: Leftover Cake

Miranda had several cakes at her party. Each cake was divided into eight equal pieces. After the party, 10 pieces were left. How many cakes is this?

Solution

Method 1: Use a Diagram

Find how many cakes are left.

Add the pieces together.

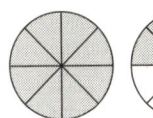

$1\frac{2}{8}$ or $1\frac{1}{4}$ cakes are left.

Method 2: Look for a Pattern

Build whole cakes.

Find a fraction of a whole that is left.

1 cake = 8 pieces

2 cakes = 16 pieces

There are 10 pieces, so there must be between 1 and 2 cakes.

There must be 1 cake and 2 pieces.

One piece is $\frac{1}{8}$ of a cake.

$1\frac{2}{8}$ or $1\frac{1}{4}$ cakes are left.

Apply

1. The school canteen sells fruit bars. One box has six bars. At the end of the day, eight bars were left. How many boxes is this?

   Complete the problem-solving plan.

   a) Find the number of _____ left.

   b) Use a diagram.
   Each rectangle represents a box of bars.
   Shade the rectangles to represent the bars left.

    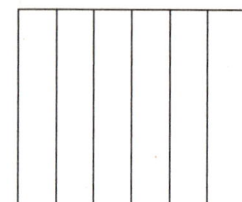

   c) Solve the problem.
   Count the number of boxes.

   There are _____ boxes left.

   d) Look for a pattern to check.
   Build whole boxes.

   1 box = _____ bars

   2 boxes = _____ bars

   3 boxes = _____ bars

   There are eight bars left, so there must be between _____ and _____ boxes.

   e) If the school bought 12 boxes, what fraction do the eight bars represent?

2. On Pizza Day, several 12-slice pizzas were ordered.

   a) After lunch, 28 slices were left. How many pizzas is this? Draw a diagram to help you.

   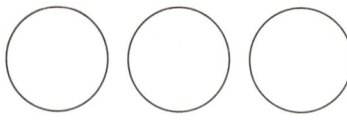

   b) If 13 pizzas were ordered, what fraction do the 28 slices represent?

3. A load of ice cream bars contains 33 boxes. Seven boxes are tiger-stripe, six are vanilla, eight are strawberry, and 12 are chocolate.

   a) What fraction of the boxes is each flavour?

   tiger-stripe: _____

   vanilla: _____

   strawberry: _____

   chocolate: _____

   b) What fraction of the boxes has tiger-stripe or vanilla?

   c) What fraction of the boxes does not have chocolate? Show or describe how you got your answer

Name: _____ Date: _____

Chapter 3: Reviewing for the Test

3.1 Add Fractions Using Manipulatives
page 86

1. Write an addition sentence to represent the fraction of each hexagon covered.

   a)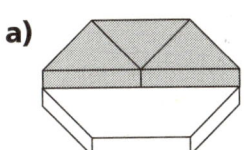

   b)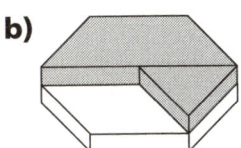

   c)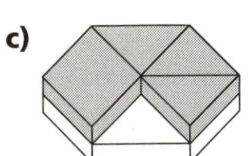

2. Complete the diagrams to show each sum.

   a) $\frac{1}{4} + \frac{1}{4} + \frac{1}{4} =$ _____

   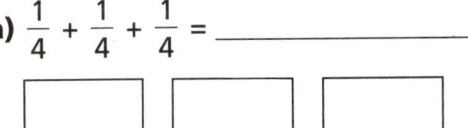

   b) $\frac{2}{3} + \frac{1}{4} =$ _____

   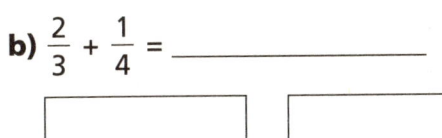

3. Write the multiplication sentence.

   $\frac{1}{4} + \frac{1}{4} + \frac{1}{4} =$ _____

3.2 Subtract Fractions Using Manipulatives
page 90

4. Write a subtraction sentence to represent each diagram. Include what remains.

   a)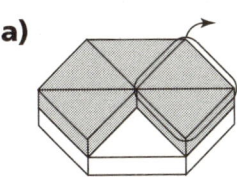

   b)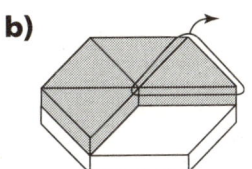

5. Draw a diagram to represent each subtraction. Include what remains.

   a) $\frac{5}{6} - \frac{1}{3} =$ _____

   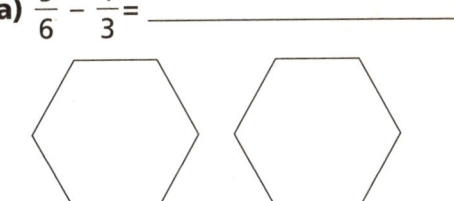

   b) $\frac{3}{4} - \frac{1}{2} =$ _____

   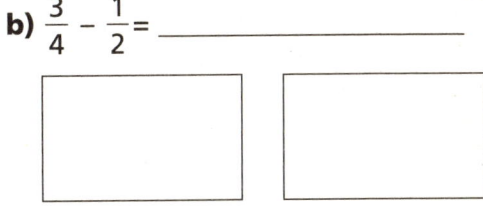

Name: _____   Date: _____

3.3 Find Common Denominators
page 94

6. Find a common denominator for each pair of diagrams. Show your work. Use multiples to check.

   a)

   b)

7. Find a common denominator for each pair of fractions. Complete the diagrams to help you. Label each rectangle with an equivalent fraction

   a) $\frac{1}{2}$ and $\frac{2}{3}$

   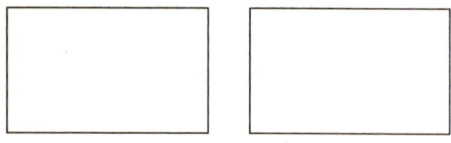

   b) $\frac{1}{6}$ and $\frac{3}{4}$

   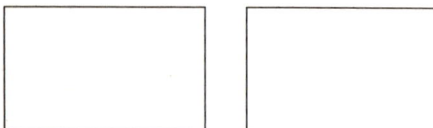

3.4 Add and Subtract Fractions Using a Common Denominator
page 98

8. Write an addition or subtraction sentence for each diagram. Then write the result.

   a)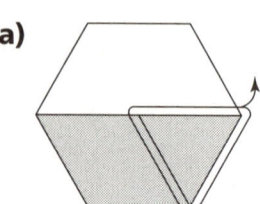

   b)

9. Complete the diagrams to represent each sentence. Draw the result.

   a) $\frac{1}{8} + \frac{1}{8} + \frac{1}{8} = $ _____

   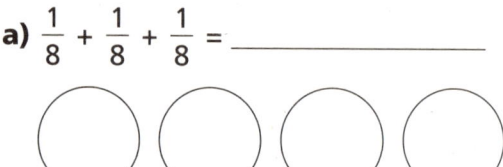

   b) $\frac{3}{4} - \frac{1}{6} = $ _____

   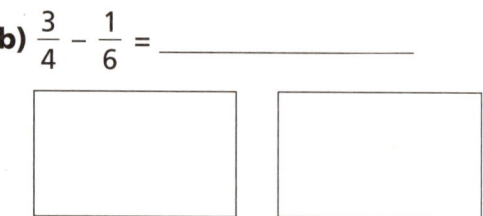

3.5 More Fraction Problems
page 104

10. A tray has apples cut into quarters. There are five pieces left. How many apples is this?

Name: _____ Date: _____

4.1 Introducing Probability

Student Text pp. 116–120

Key Ideas Review

Fill in the blanks with words from the list.

| All | experiment | outcomes | fraction |
| Probabilities | decimal | favourable | |

1. The _____ of an experiment are the possible results.

2. _____ can be estimated from repeated trials of an _____.

3. Probabilities can also be calculated.

   Probability = $\dfrac{\underline{\hspace{2em}} \text{ outcomes}}{\underline{\hspace{2em}} \text{ outcomes}}$

4. Probabilities can be shown as a _____ or a _____.

Example: Calculate Probability by Outcomes

There are 10 coloured tiles in a bag: 4 yellow, 3 blue, 2 black, and 1 green.

a) How many possible outcomes are there for picking one tile out of the bag?

b) If you want to pick a blue tile, how many favourable outcomes are there?

c) What is the probability of picking a blue tile at random?
   Express your answer as a fraction and as a decimal.

Solution

a) There are 10 tiles in the bag. So, there are 10 possible outcomes.

b) There are 3 blue tiles, so there are 3 favourable outcomes.

c) Probability (blue tile) = $\dfrac{\text{favourable outcomes}}{\text{all outcomes}}$

   $= \dfrac{\text{number of blue tiles}}{\text{total number of tiles}}$

   $= \dfrac{3}{10}$

   $= 0.3$

The probability of picking a blue tile is $\dfrac{3}{10}$, or 0.3.

Name: _____ Date: _____

Practise

1. In each situation, state the total number of outcomes and the number of favourable outcomes. Then write the probability as a fraction.

   a) Spinning a 7.

   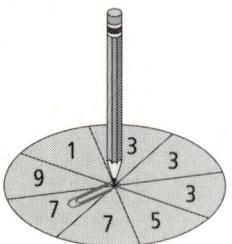

   b) Drawing a rectangle.

   Total number of outcomes = _____

   Number of favourable outcomes = _____

   Probability (7) = _____

   Total number of outcomes = _____

   Number of favourable outcomes = _____

   Probability (rectangle) = _____

2. Rhonda has a drawer containing the following numbers of coloured barrettes:
   - 3 green (G)
   - 4 black (B)
   - 5 white (W)
   - 2 pink (P)

   What is the probability of randomly picking each colour of barrette?

   Probability (green) = _____   Probability (black) = _____

   Probability (white) = _____   Probability (pink) = _____

3. Rhonda pulls out two barrettes at random.

   a) Create a model for drawing barrettes.

   b) Simulate drawing two barrettes at the same time. Record whether the two are the same colour or not.

   c) Complete the tally chart by drawing 50 pairs.

	Tally	Frequency
Matched		
Unmatched		
	Total Trials	

   d) According to your trials, what is the probability that Rhonda selects matching barrettes?

   $\dfrac{\text{matched pairs}}{\text{total trials}}$ = _____

4.2 Organize Outcomes

Student Text pp. 121–125

Key Ideas Review

The key word spinner has three possible outcomes.
Fill in the blanks with the word(s) that match the probabilities.
Tree diagrams and lists can ...

- be used to determine _____ $\left(\frac{2}{6}\right)$

- help to organize the _____ $\left(\frac{3}{6}\right)$ of a probability experiment

- be created by modelling with _____ $\left(\frac{1}{6}\right)$ objects

Example 1: Use a List or Model

Create an organizer, with probabilities, for a spinner with equal orange, purple, and green sectors.

Solution

Method 1: Make a List

- Spinning orange: Probability = $\frac{1}{3}$
- Spinning purple: Probability = $\frac{1}{3}$
- Spinning green: Probability = $\frac{1}{3}$

Method 2: Test the Spinner

Make and test the spinner. You should find that the three outcomes are equally likely.

Probability (orange) = $\frac{1}{3}$

Probability (purple) = $\frac{1}{3}$

Probability (green) = $\frac{1}{3}$

Example 2: Use a Tree Diagram

a) Create a tree diagram to show the possible outcomes from flipping two coins.

Solution

a)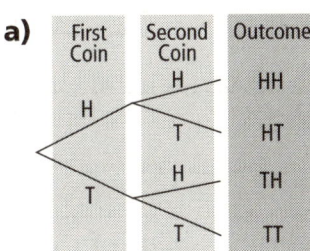

b) What is the probability of no heads?

c) What is the probability of exactly one tail?

b) There is only one favourable outcome, (T, T).

Probability (no heads) = $\frac{\text{favourable outcomes}}{\text{all outcomes}}$

$= \frac{1}{4}$

c) There are two favourable outcomes, (H, T) and (T, H).

Probability (one tail) = $\frac{\text{favourable outcomes}}{\text{all outcomes}}$

$= \frac{2}{4} = \frac{1}{2}$

Practise

1. Create an organized list, with probabilities, for this spinner.

_____

_____

_____

_____

_____

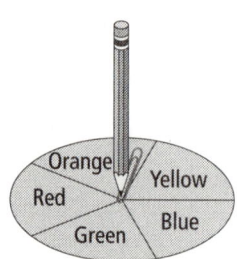

2. Alexi has labelled tokens in two bags as shown. He picks one token from each bag.

a) Create a tree diagram for this situation.

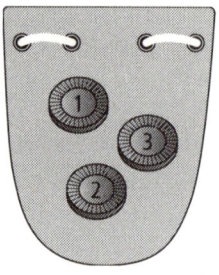 

b) How many possible outcomes are there?

_____

c) What is the probability of choosing the 2 and the 12?

Probability (2 and 12) = _____

d) What is the probability of choosing two odd numbers?

Probability (two odd numbers) = _____

Literacy Connections

Read tree diagrams from left to right.
- The branches on the left of the tree show the outcomes for one object.
- The branches on the right show the outcomes for the other object.
- The far right of the diagram lists the combined outcomes.

Tree diagrams can also be created for more than two objects.

Name: _____ Date: _____

4.3 Use Outcomes to Predict Probabilities

Student Text pp. 126–130

Example 1: Probability in Number Cards

Students play a game with 8 cards as shown.
Determine the probability of each draw from the deck.

a) any card with a ■

b) any card with an odd number

Solution

a) **Method 1: Count Outcomes**
There are two cards with a ■. The probability of picking a ■ card is $\frac{2}{8}$ or $\frac{1}{4}$.

Method 2: Use Proportions
Exactly one-quarter of the deck has a ■. The probability of picking a ■ card is $\frac{1}{4}$.

b) **Method 1: Count Outcomes**
There are four cards that are favourable outcomes. The probability of picking a card with an odd number is $\frac{4}{8} = \frac{1}{2}$.

Method 2: Use Logic
Exactly half of the cards have an odd number on them. The probability of picking a card with an odd number is $\frac{1}{2}$.

Example 2: Outcomes of Multiple Spins

In a board game, Silvio spins two spinners. He gets to move his piece ahead if he spins two odd numbers. What is the probability that Silvio gets to advance his piece?

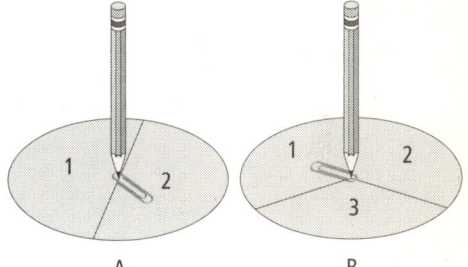

A   B

Solution

Step 1: Write lists for the outcomes of each spinner.

Spinner A: • spinning 1: $\frac{1}{2}$

• spinning 2: $\frac{1}{2}$

Spinner B: • spinning 1: $\frac{1}{3}$

• spinning 2: $\frac{1}{3}$

• spinning 3: $\frac{1}{3}$

Step 2: Draw a tree diagram to show the possible outcomes.

There are two favourable outcomes out of a total of six outcomes.

Spinner A	Spinner B	Outcome
1	1	1, 1
	2	1, 2
	3	1, 3
2	1	2, 1
	2	2, 2
	3	2, 3

Probability (Silvio moves) = $\frac{2}{6} = \frac{1}{3}$.

Name: _____ Date: _____

Practise

1. Using the cards shown, calculate the probabilities to move through the grid. Draw a line to each answer from the previous answer to create a picture.

   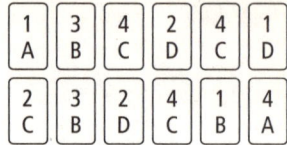

   a) Drawing a 3.
   b) Drawing a C.
   c) Drawing a 1.
   d) Drawing a 2 and a D.
   e) Drawing an odd number.
   f) Drawing an odd number and a vowel.
   g) Drawing a 4 and a C.
   h) Drawing an A or a B.

$\frac{7}{12}$	$\frac{4}{12}$	$\frac{11}{12}$
$\frac{2}{12}$	$\frac{12}{12}$	$\frac{3}{12}$
$\frac{8}{12}$	$\frac{6}{12}$	$\frac{10}{12}$
$\frac{1}{12}$	$\frac{9}{12}$	$\frac{5}{12}$

   Start

2. What are two ways to determine the probability of drawing a C?

3. Yasmina is on a game show. If she spins PRIZE on both spinners, she wins. Use lists and a tree diagram to determine the probability that Yasmina will win.

   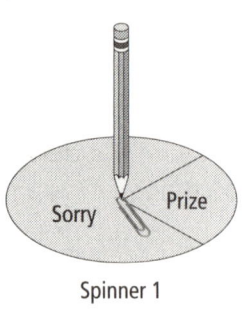
   Spinner 1

   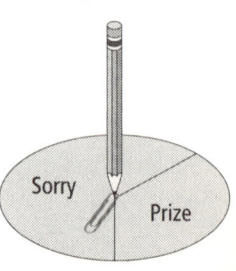
   Spinner 2

   Probability (winning, both PRIZE) = _____

44 MHR • Chapter 4: Probability and Number Sense

Name: _____   Date: _____

4.4 Extension: Simulations

Student Text pp. 131–133

Key Ideas Review

Circle the correct bolded word(s) to complete the statements.

1. A **simulation / situation** is an experiment that can be used to model a real situation involving **frequency tables / probabilities**.

2. There are **three / many** different ways to simulate a situation.

Practise

1. Describe a method that could be used to simulate each of the following situations. Explain your choices.

    a) Randomly choosing one vacation destination from a list of six choices.

    b) Randomly picking one movie to rent from a selection of 10.

Apply

2. Evinna works at a clothing store. On Thursday she sold 30 items: 15 T-shirts, 6 pairs of jeans, 6 sweatshirts, and 3 pairs of shorts. Describe, create, and conduct a simulation based on this situation. Run your simulation 50 times and record the results in the tally chart.

Clothing Item	Tally
T-shirt	
Jeans	
Sweatshirt	
Shorts	

4.5 Apply Probability in Sports and Games

Student Text pp. 134–139

Key Ideas Review

Complete the statements. Unscramble the letters to fill in the blanks.

1. Knowledge of _____ can help in sports and games.
   BROITABILPY

   Strategies that give more favourable _____ improve
   TSOOCEUM

   your chances of winning.

2. Batting _____ in baseball and goals scored in hockey or
   EGARESAV

   soccer are an indication of an athlete's _____.
   PCOMFEARNER

Fill in the blanks using the numbers 276 and 1000.

3. A batting average can be written in two ways: 0.276 or $\frac{}{}$.

   This means that out of every _____ times at bat, this player had
   _____ hits.

Example: Probability in Sports

Rick is a kicker for a champion football team, the Bulldogs.

He has a kicking average of 0.800.

a) Estimate the probability of Rick kicking a field goal.

b) Estimate the probability of Rick not kicking a field goal.

c) How many successful field goals will Rick probably kick in his next 40 attempts?

Solution

a) A kicking average of 0.800 means that Rick kicks a successful field goal 800 times in 1000 attempts. So, the estimated probability of him kicking a field goal is $\frac{800}{1000}$, or $\frac{4}{5}$.

b) Out of every 1000 attempts, Rick is expected not to kick a field goal 200 times. The estimated probability of him not kicking a field goal is $\frac{200}{1000}$, or $\frac{1}{5}$.

c) **Method 1: Use Fraction Strips**

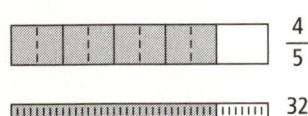 $\dfrac{4}{5}$

$\dfrac{32}{40}$

Method 2: Use Equivalent Fractions

$$\dfrac{4}{5} = \dfrac{4 \times 8}{5 \times 8} = \dfrac{32}{40}$$

In 40 attempts, Rick will probably kick 32 field goals.

Practise

1. Ruben and Ang are playing *Rock, Paper, Scissors*. Use the tree diagram to answer the questions.

   a) Number of outcomes = _____

   b) Probability (tie) = _____

   c) Probability (Ruben wins) = _____

   d) Probability (Ang wins) = _____

   e) How could you have calculated the probability of a tie without a tree diagram?

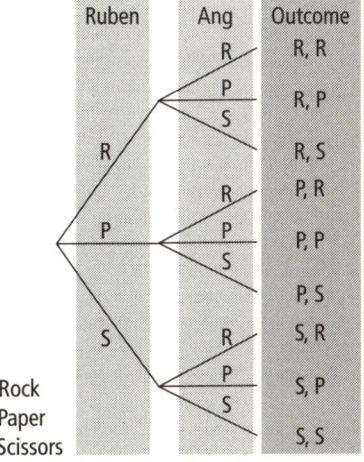

R = Rock
P = Paper
S = Scissors

2. Use a tree diagram to show the probability of drawing two cards the same.

   Probability (two the same) = _____

Apply

3. Matt's hockey team has a winning average of 0.600.

   a) Estimate the probability of Matt's team winning its next game.

      Probability (win) = _____

   b) Estimate the probability of Matt's team losing its next game, as a fraction and a decimal.

      Probability (not winning) = _____

   c) How many wins, on average, will Matt's team get in its next 20 games?

      Number of wins = _____

4.5 Apply Probability in Sports and Games • MHR 47

Chapter 4: Reviewing for the Test

4.1 Introducing Probability
page 116

1. A tally chart was made to record the results of rolling a die.

	Tally	Frequency
Odd Number	|||| |||| |||| |||| |||	
Even Number	|||| |||| |||| |||| |||| ||	
	Total Trials	

a) Complete the tally chart and frequency table.

b) What fraction of the rolls resulted in an odd number? _____

c) What is the actual probability of rolling an odd number? _____

2. A bag contains 30 tokens. There are 20 S-tokens, 8 M-tokens, and 2 L-tokens.

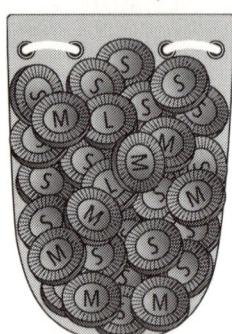

What is the probability of picking each letter? Write your answers as fractions and as decimals.

a) Probability (L) = _____

b) Probability (M) = _____

c) Probability (S) = _____

4.2 Organize Outcomes
page 121

3. Create an organized list with probabilities for this spinner.

_____

_____

_____

_____

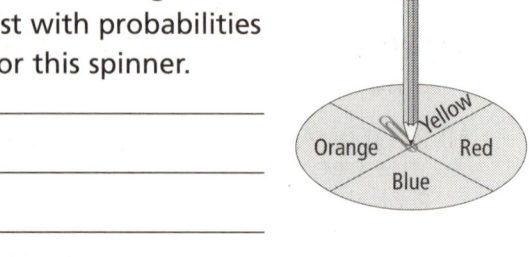

4. Lateesha is playing a board game. She adds the total of the two spinners, then moves that many spaces.

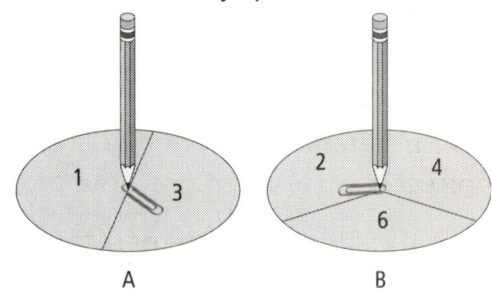

a) Create a tree diagram for this situation.

b) What is the probability that Lateesha moves five places?

Probability (5) = _____

Name: _____   Date: _____

4.3 Use Outcomes to Predict Probabilities
page 126

5. A letter is drawn at random from the word OCEAN. Then another is drawn from the word SEA.

   a) Organize the outcomes of the two draws.

   b) What is the probability of not picking S?

   Probability (no S) = _____

   c) What is the probability of at least one letter being a vowel?

   Probability (at least one vowel) = ____

   d) What is the probability of both letters being vowels?

   Probability (both vowels) = _____

6. Which statements about dice are correct?

   A. You can't roll 20 with only three dice.

   B. With one die, you are more likely to roll a number less than 6.

   C. With two dice, you are more likely to roll a number less than 6.

   D. There are four ways to roll 7 with two dice.

4.4 Extension: Simulations
page 131

7. a) Describe a simulation for rolling a cube labelled A, A, B, B, C, D.

   b) What is the probability of not rolling a B

   Probability (no B) = _____

4.5 Apply Probability in Sports and Games
page 134

8. Eric has a batting average in baseball of 0.300. The batting average of his teammate Juan is 0.200.

   a) Determine the probability of each player getting a hit. Write it as a fraction.

   Eric: _____   Juan: _____

   b) How many hits on average will Eric get in 20 at-bats?

   Hits = _____

   c) How many hits on average will Juan get in 30 at-bats?

   Hits = _____

9. Michelle plays basketball for her school team. She averages 7 baskets in every 10 attempts.

   a) Last game, she scored 12 baskets on 15 attempts. How did she do compared to her average?

   b) What is the probability that she will make her first shot next game?

Name: _____ Date: _____

5.1 Fractions and Decimals

Student Text pp. 152–157

Key Ideas Review

Use the following to fill in the blanks.

> denominator   hundred   hundredth   numerator

Convert a fraction to a decimal.

A. _____

$\frac{7}{9}$ → 7 ÷ 9 = 0.777 ...

B. _____ = $0.\overline{7}$

Convert a decimal to a fraction.

$0.65 = \frac{65}{100}$ D. _____

C. _____ = $\frac{13}{20}$

Example 1: Compare and Order Fractions

Dessert Cafe sells cake by the slice. At the end of a day, $2\frac{3}{5}$ carrot cakes, $2\frac{3}{4}$ chocolate cakes, and $2\frac{5}{6}$ vanilla cakes had been sold.

a) Represent the amounts with a diagram.

b) Write each amount as a decimal.

c) Which cake was the most popular? How do you know?

Solution

a) carrot:

chocolate:

b) $2\frac{3}{5} = 2 + 3 ÷ 5$

$2\frac{3}{5} = 2 + 0.6$

$2\frac{3}{5} = 2.6$

$2\frac{3}{4} = 2 + 3 ÷ 4$

$2\frac{3}{4} = 2 + 0.75$

$2\frac{3}{4} = 2.75$

vanilla: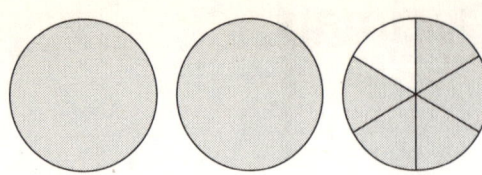

$2\frac{5}{6} = 2 + 5 \div 6$ $5 \div 6 = 0.83333333$

$2\frac{5}{6} = 2 + 0.8333\ldots$

$2\frac{5}{6} = 2 + 0.8\overline{3}$

$2\frac{5}{6} = 2.8\overline{3}$

c) $2.8\overline{3} > 2.75 > 2.6$
The cakes, from the most popular to the least, are vanilla, chocolate, and carrot.

Example 2: Convert Decimals to Fractions

A recipe calls for 0.56 kg of flour. Convert 0.56 kg into a fraction.

Solution

$0.56 = \frac{56}{100}$

0.56 kg is $\frac{56}{100}$ of a kilogram.

You can also write the fraction as $\frac{14}{25}$ of a kilogram.

$\frac{56}{100} = \frac{56 \div 4}{100 \div 4}$

$\frac{56}{100} = \frac{14}{25}$

Practise

1. What fraction is represented by the shaded area?

   a) 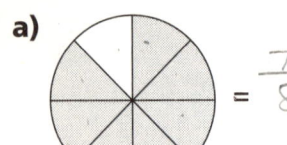 $= \frac{7}{8}$

   b) 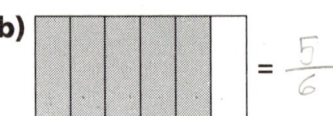 $= \frac{5}{6}$

   c) $= 2\frac{2}{3}$

2. Write each fraction in question 1 as a decimal.

   a) $7 \div 8 = 0.87$

   b) $5 \div 6 = 0.83$

   c) $\frac{2}{3}$ $2 \div 3 = 0.66$
   2.66

3. Order the numbers in question 2 from least to greatest.

   $0.83 < 0.87 < 2.66$

4. Express each decimal as a fraction. Simplify the fraction.

   a) 0.25 $\frac{25}{100}$

   b) 0.3 $\frac{3}{10}$

   c) 0.58 $\frac{58}{100}$

Name: _____ Date: _____

5.2 Calculate Percents

Student Text pp. 158–161

Key Ideas Review

1. These are the instructions for writing a fraction as a percent. What operations are missing?

    numerator $\frac{\div}{\ }$ denominator $\underline{\times}$ 100%

2. To show a percent, you can use diagrams. What percent is shown in each diagram?

    a) 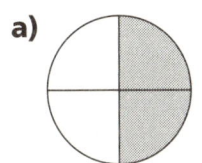 50%

    b)  50%

    c) 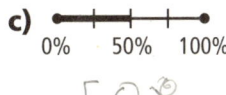 50%

Example: Write Fractions and Decimals as Percents

Flora Stickers claims that at least 80% of the stickers in each of its packages are flowers. The rest are leaves.

Jeannie counts 15 flowers out of 18 stickers. Is that fraction at least 80%?

Solution

Method 1: Convert to Decimal, then to Percent

$\frac{15}{18} = 0.8333...$

$= 0.8\overline{3} \times 100\%$

$\doteq 83\%$

More than 80% of Jeannie's stickers are flowers.

Method 2: Convert to Percent

$\frac{15}{18}$ of $100\% = \frac{15}{18} \times 100\%$

$= \frac{1500}{18} \%$

$\doteq 83\%$

Practise

1. For a survey, 50 students were asked, "What is your favourite colour?"

    a) The results are shown. Complete the chart.

Colour	Number	Decimal	Percent
Blue (B)	15/50	0.3	30%
Red (R)	10/50	0.2	20%
Black (Bl)	20/50	0.4	40%
Green (G)	5/50	0.1	10%

    b) Complete each visual to show the results.

    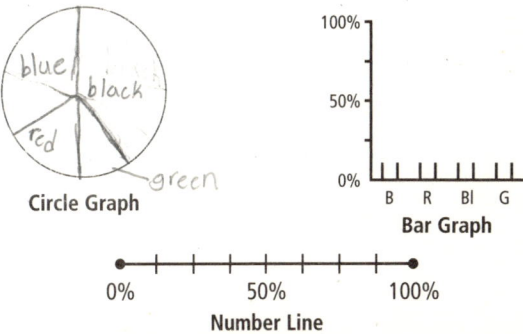

    Circle Graph

    Bar Graph

    Number Line

5.2 Calculate Percents • MHR 53

5.3 Fractions, Decimals, and Percents

Student Text pp. 162–165

Key Ideas Review

Match the fractions with the correct decimals. Use their matching letters to spell the missing word.

0.25	0.6	0.5	0.7	0.43	0.75	0.1	0.85	0.33	0.3	0.9	0.55	0.4	0.12	0.21
A	C	D	E	H	I	L	M	N	O	R	S	T	V	Y

1. $\dfrac{1}{2} = $ __D__
2. $\dfrac{30}{40} = $ __I__
3. $\dfrac{3}{25} = $ __V__
4. $\dfrac{6}{8} = $ __I__
5. $\dfrac{6}{12} = $ __D__
6. $\dfrac{7}{10} = $ __E__

To express a percent as a fraction, __D__ __i__ __v__ __i__ __d__ __e__ the number by 100.
                                         1 2 3 4 5 6

Example: Express Percents as Fractions and Decimals

Geoff and Lindsay both ran for treasurer of the Social Council. The results are shown in the table.

Candidate	Percent
Geoff	55%
Lindsay	45%

a) What fraction of the votes did each person receive? What is the decimal equivalent?

b) Of the 200 votes cast, how many votes did each person get?

Solution

a)

Candidate	Percent	Fraction	Decimal
Geoff	55%	$\dfrac{55}{100} = \dfrac{11}{20}$	$\dfrac{11}{20} = 0.55$
Lindsay	45%	$\dfrac{45}{100} = \dfrac{9}{20}$	$\dfrac{9}{20} = 0.45$

You could also use 55% = 55 ÷ 100
                           = 0.55

You could also use 45% = 45 ÷ 100
                           = 0.45

b) Geoff:
Number of votes = 55% × Total number of votes
= 0.55 × Total number of votes
= 0.55 × 200
= 110
Geoff received 110 votes.

Lindsay:
Number of votes = 45% × Total number of votes
= 0.45 × Total number of votes
= 0.45 × 200
= 90
Lindsay received 90 votes.

Practise

1. Complete the table. Round your answers to the nearest percent.

Fraction	Decimal	Percent
$\frac{2}{7}$	0.28	28%
$\frac{45}{100}$	0.45	45%
$\frac{85}{125}$	0.68	68%
$\frac{22}{100}$	0.22	22%

Apply

2. The table shows the records of several teams in a hockey league.

   a) Complete the table. Express each team's winning record as a fraction of its games played.

Team	Games Played	Games Won	Games Won / Games Played	Decimal Equivalent of Fraction	Order
Tigers	8	3	$\frac{3}{8}$	0.37	37%
Lions	6	4	$\frac{4}{6}$	0.66	66%
Pumas	12	6	$\frac{6}{12}$	0.5	50%
Jaguars	8	5	$\frac{5}{8}$	0.62	62%
Leopards	12	5	$\frac{5}{12}$	0.41	41%

   b) Use the decimal equivalents of the fractions to rank the teams from 1 (best) to 5 (worst).

   Which team had the best record? __Lions__

   c) If the Lions played nine more games, how many would they be expected to win? __10__

   Hint
   Figure out what fraction of their games the Lions tend to win.

5.4 Apply Fractions, Decimals, and Percents

Student Text pp. 166–171

Example: Compare Sports Data

Savitri, Jose, and Tanya are the best players on the YMV volleyball team.

Based on the statistics, who is the best server? Round your answer to the nearest percent.

Player	Serves Attempted	In (✓) / Out (✗)
Savitri	13	✓✓✓✗✓✓✗✓✓✗✗✓✓
Jose	17	✓✓✓✗✗✗✓✓✗✗✓✓✓✗✗✓
Tanya	11	✓✓✓✓✗✓✗✗✓✓✓

Solution

Savitri:

9 serves out of 13 attempts.

$\frac{9}{13} = 0.692...$

Express this as a percent:
0.692... = 0.692... × 100%
= 69.2...%
≐ 69%

Jose:

10 serves out of 17 attempts.

$\frac{10}{17} = 0.588...$

Express this as a percent:
0.588... = 0.588... × 100%
= 58.8...%
≐ 59%

Tanya:

8 serves out of 11 attempts.

$\frac{8}{11} = 0.7272...$

Express this as a percent:
$0.\overline{72} = 0.\overline{72} \times 100\%$
= 72.72...%
≐ 73%

Tanya was successful in about 73% of her serves, so Tanya is the best server.

Apply

1. The 25 students in Theo's class were asked, "What is your favourite flavour of ice cream?" The results are shown.

Flavour	Number of Students		Fraction	Decimal
Chocolate	|||| ||||	10	$\frac{10}{25}$	
Strawberry	|||| ||	7		
Vanilla	|||| |||	8		

a) What fraction of the class liked each flavour best? Fill in the table.

b) What percent of the class liked each flavour best? Fill in the table.

c) Use the circle graph and label it to show the percents.

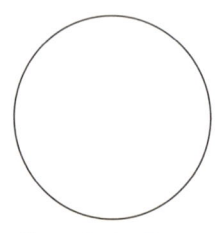

Favourite Ice Cream

Making Connections

Record the amounts of time you spent on certain activities in one day.

Some activities are: watching TV, homework, school, eating, playing/talking with your friends, and so on.

Calculate what percent of one day each activity took. Use a diagram to show your results.

Chapter 5: Reviewing for the Test

5.1 Fractions and Decimals
page 152

1. Place each fraction on the number line. Place the number labels under the line as shown.

   a) $\frac{3}{4}$ b) $\frac{1}{4}$ c) $\frac{8}{9}$

   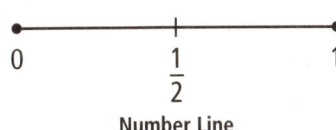
   Number Line

2. Convert each fraction in question 1 to a decimal. Place each decimal above its fraction on the number line.

3. Maggie's three dogs are fed the same amount of dog food at each meal.
   - Tipper ate $\frac{3}{5}$ of her meal.
   - Fudge ate $\frac{3}{4}$ of his meal.
   - Zip ate $\frac{5}{6}$ of his meal.

   Who ate the most of their meal? Draw diagrams to help you explain.

5.2 Calculate Percents
page 158

4. Lia's scores on her last three math assignments are $\frac{21}{25}$, $\frac{37}{40}$, and $\frac{53}{60}$.

   a) Write each score in the table as a decimal.

   b) Write each score in the table as a percent.

	Fraction	Decimal	Percent
Assignment 1	$\frac{21}{25}$		
Assignment 2	$\frac{37}{40}$		
Assignment 3	$\frac{53}{60}$		

   c) On which assignment did Lia achieve the highest grade? _____

5. A record company claims that of all the CDs produced, only 0.15% will be defective. Of 55 000 CDs produced, 100 were defective. Is the company right? Explain.

5.3 Fractions, Decimals, and Percents
page 162

6. Complete the table. Round decimals to the nearest tenth. Round percents to the nearest whole number.

Fraction	Decimal	Percent
	0.7	
		55%
	0.375	
$\frac{13}{55}$		

7. What's wrong? Ming wrote 0.8 as a fraction this way.

$0.8 = 8\%$
$= \frac{8}{100}$
$= \frac{2}{25}$

Correct Ming's mistake. Explain your answer.

5.4 Apply Fractions, Decimals, and Percents
page 166

8. A CD cover is made up of two colours.

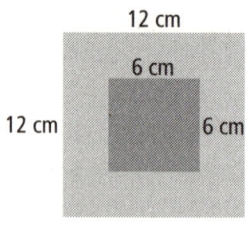

a) What percent of the cover is taken up by the inner square?

b) What percent of the cover is taken up by the outer square?

9. a) What is 40% of $1.50?

b) Use your answer to figure out $\frac{3}{5}$ of $1.50.

10. Khan can make a successful volleyball serve 8 times out of 10. Raj can do the same 12 times out of 14.

a) Show each player's statistic as a fraction, a decimal, and a percent.

Khan: **Raj**

fraction =   fraction =

decimal =   decimal =

percent =   percent =

b) Who is the better server? Explain.

Name: _____   Date: _____

6.1 / 6.2 Investigate and Describe Patterns
Organize, Extend, and Make Predictions

Student Text pp. 180–189

Key Ideas Review

Write the next terms in each pattern. The answers will spell the missing word.

- In a pattern, you can _____ what comes next.
- Some patterns are based on **number operations**. Other patterns are based on geometric shapes.
- To describe a pattern:
  - Identify the **first term**.
  - Describe how the items that follow are **generated**.
  - Relate each new item to previous items or to the counting numbers.
- Any letter can be used as a **variable** to represent a number.
- A variable can be used in an **expression** that shows how a pattern works.

H, J, L, N, ___
N, O, P, Q, ___
A, B, C, D, ___
L, J, H, F, ___
M, L, K, J, ___
K, I, G, E, ___
H, K, N, Q, ___

Example: Describe and Predict

Examine this pattern: 5, 9, 13, 17, ...

a) Describe the pattern in words.

b) Describe the pattern using a variable.

c) What are the next three numbers in the pattern?

Solution

a) The second number, 9, is 4 more than the first number, 5.
The next number, 13, is 4 more than 9.
The pattern starts at 5 and each number is 4 more than the one before it.

b) Each number is 1 more than a multiple of 4.
If n is the variable, each number is $4 \times n + 1$, or $4n + 1$ for short.

c) **Method 1: Complete the Sequence**
Since each number is 4 more than the one before it, the next number is $17 + 4$, which is 21.
Next are $21 + 4 = 25$ and $25 + 4 = 29$.

Method 2: Use Equations
Each number is $4n + 1$.
The fifth number is $4 \times 5 + 1 = 21$.
The sixth number is $4 \times 6 + 1 = 25$.
The seventh number is $4 \times 7 + 1 = 29$.

Practise

1. Marg was told to look for a pattern in these numbers: 45, 40, 35, 30, ...

   a) Does the pattern involve addition, subtraction, multiplication, or division?

   b) How does each number relate to the one before it?

   c) Complete this expression for terms in the pattern. 50 – ___ x

   d) Use the expression to calculate the next three numbers in the pattern.

2. Describe each pattern in words.

   a) 7, 21, 63, 189, ... _____

   b) 1, 11, 111, 1111, ... _____

   c) /, ⁊, ⁊⁊, ⁊⁊⁊, ... _____

3. Describe each pattern in words and using a variable.

   a) 6, 12, 18, 24, 30, ... _____

   _____

   b) 5, 8, 11, 14, 17, ... _____

   _____

   c) 42, 39, 36, 33, 30, ... _____

   _____

4. What are the next three numbers in each pattern in question 3?

   a)

   b)

   c)

5. Complete each pattern with the appropriate value.

   a) 2, 4, _____, 8, _____, 12, ...

   b) _____, 6, 12, _____, 48, 96, ...

   c) 26, _____, 20, 17, _____, 11, ...

60   MHR • Chapter 6: Patterning

6.3 / 6.4 Explore Patterns on a Grid or in a Table of Values / Express Simple Relationships

Student Text pp. 190–199

Key Ideas Review

Find the bolded words in the puzzle. Unscramble the leftover letters to spell the missing word.

- Patterns can be shown by listing ordered pairs in a **table** of **values**.
- Patterns can also be **plotted** on a coordinate **grid**.
- A pattern **rule** is a description of a pattern, in **words**. It is often used to **predict** the pattern.
- A **pattern** between two sets of numbers is called a relationship.
- Ordered pairs of **data** can be analysed to identify a relationship.
- You can use an algebraic _____ to express a relationship.

N	E	T	P	E	N	V
T	R	W	L	L	G	A
A	D	E	O	U	R	L
B	Q	A	T	R	I	U
L	U	I	T	T	D	E
E	A	O	E	A	A	S
T	C	I	D	E	R	P

Example: Describe an Equation and Plot Points

Matt is calculating the surface areas of prisms with the same base. He counts 16, 24, and 32 squares for the surface areas of these first three shapes, but then he runs out of cubes.

a) How can Matt express the numbers algebraically?

b) Write the first five terms in a table of values.

c) Plot these values on a graph. Include the values as ordered pairs.

d) What do you notice about the points? What do you think the next point will be?

Solution

a) The numbers are increasing by 8. Each number is a multiple of 8. If d is the height of the prism, then the surface area is $8 + 8d$.

b)

d	1	2	3	4	5
$8 + 8d$	16	24	32	40	48

Name: Zain

Date:

c)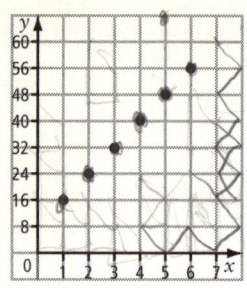

d) All of these points are in a straight line that is increasing. Use a ruler to see where the line reaches 6. The next point should be at (6, 56).

Practise

1. Tate plans to buy five individual hockey cards and a few packs of cards. Each pack contains eight cards.

a) How many cards will Tate have if he buys

Two packs of cards? ___21___

Only one pack? ___13___

b) If p is the number of packs he buys, write an expression for the total number of cards Tate buys using p.

c) Complete this table of values.

Number of packs	Number of cards
0	5
1	13
2	21
3	29
4	37

d) Plot these values as points.

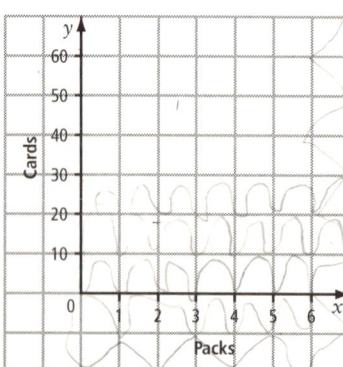

62   MHR • Chapter 6: Patterning

e) Describe the pattern on the graph.

2. Complete this table of values.

n	3n	2n + 1	100 − 15n	11 + 8n
0	0	1	100	11
1	3	3	85	19
2	6	5	70	27
3	9	7	55	35
4	12	9	40	43

3. Using question 2, plot the first five points for 3n and for 2n + 1.

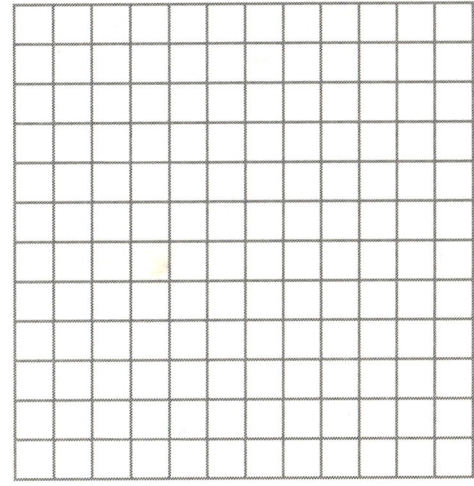

4. Describe the pattern on this graph. What might be the pattern rule for the y-axis values?

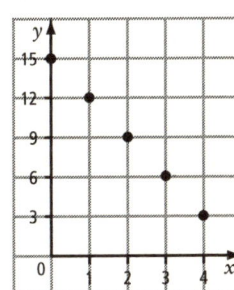

Name: _____ Date: _____

Chapter 6: Reviewing for the Test

6.1 **Investigate and Describe Patterns**
6.2 **Organize, Extend, and Make Predictions**

page 180

1. **a)** How would you find the next term in this sequence? 96, 87, 78, 69, ... Circle the answer.

   A subtract 11   **B** subtract 9

   C divide by 2   **D** divide by 11

   b) What are the next two terms?

   _____   _____

2. What is the next term in each pattern?

   a) 7, 14, 28, 56, _____

   b) 80, 120, 160, 200, _____

   c) X, U, R, O, _____

   d) $\frac{7}{32}, \frac{7}{16}, \frac{7}{8}, \frac{7}{4},$ _____

3. A tower is made of connecting blocks. Each level has half as many blocks as the one below it. The second level has 240 blocks.

   How many blocks are on each level?

   Top : _____

   Fourth: _____

   Third : _____

   Second: **240**

   Bottom: _____

6.3 **Explore Patterns on a Grid or in a Table of Values**
6.4 **Express Simple Relationships**

page 190

4. Complete this table of values.

n	5n	3n + 2	50 + 7n	50 − 7n
0				
1				
2				
3				
4				

5. **a)** Plot the first five terms of $3n + 2$ on this graph.

   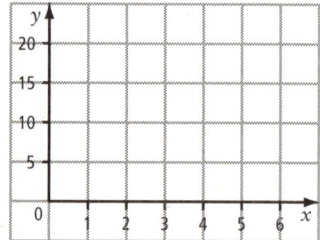

   b) Describe the pattern of these points.

   ### Making Connections

   Patterns can be found in nature.

   One example is the pattern of the seeds in a sunflower. Find a close-up picture of the flower. What is the pattern? Where else in nature do you see a pattern?

7.1 Understand Exponents

Student Text pp. 210–213

Key Ideas Review

Fill in the blanks with words from the list.

> two   multiplication   three

1. Exponents represent repeated _____.

2. A square is the product of _____ equal factors.

3. A cube is the product of _____ equal factors.

base
- the factor you multiply

exponent
- the number of factors you multiply

exponential form
- a shorter method for writing numbers expressed as repeated multiplication
  $4 \times 4 \times 4 = 4^3$

Example 1: Evaluate Squares

Find the area of a square with each side length.
a) 5   b) 4.3

Solution

a) $A = s^2$
$A = 5^2$
$A = 5 \times 5$
$A = 25$
The area is 25 square units.

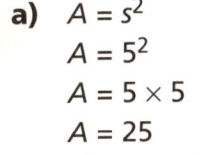

b) Use a calculator.
$A = s^2$
$A = 4.3^2$
$A = 18.49$
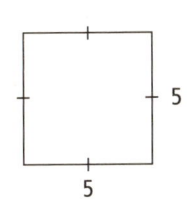
The area is 18.49 square units.

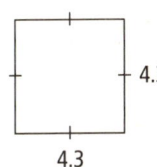

Example 2: Evaluate Cubes

Find the volume of a cube with each edge length.
a) 7   b) 12.5

Solution

a) $V = l^3$
$V = 7^3$
$V = 7 \times 7 \times 7$
$V = 49 \times 7$
$V = 343$
The volume is 343 cubic units.

b) Use a calculator.
$V = l^3$
$V = 12.5^3$
$V = 1953.125$
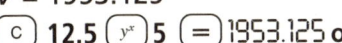

The volume is 1953.125 cubic units.

Name: _____   Date: _____

Example 3: Use Exponential Form

Write each repeated multiplication in exponential form.

a) $14 \times 14 \times 14$

b) 12×12

Solution

a) $14 \times 14 \times 14 = 14^3$
There are three 14s, so the exponent is 3.

b) $12 \times 12 = 12^2$
There are two 12s, so the exponent is 2.

Practise

1. Complete the chart.

Length of Edge (cm)	Area of Square (cm²)	Volume of Cube (cm³)
3		
10		
2		
6		
2.5		

2. Write each repeated multiplication in exponential form.

a) $10 \times 10 \times 10 \times 10 =$

b) $2 \times 2 \times 2 \times 2 \times 2 \times 2 \times 2 \times 2 =$

c) $6 \times 6 \times 6 \times 6 \times 6 =$

d) $73 \times 73 \times 73 \times 73 \times 73 \times 73 =$

3. Write each as a repeated multiplication.

a) $7^3 =$

b) $5^4 =$

c) $16^2 =$

d) $6^7 =$

Apply

4. Jason has a cubic storage box with each edge having a length of 9 cm.

   What is the volume of his treasure chest?

   Volume =

5. Write these expressions in order from least to greatest. The rearranged letters will answer the riddle.

   What did one magnet say to the other magnet?

   2^2 15 4^2 2^3 3^2 10   5 1^3 20
   T   T   E   R   A   C   T   A   D

   "I feel ___ ___ ___ ___ ___ ___ ___ ___ ___ to you!"

7.2 Represent and Evaluate Square Roots

Student Text pp. 214–217

Key Ideas Review

Unscramble the letters to complete the sentences.

1. The _____ _____ of a square represents the square root of the area.
   DSEI   EGLNHT

2. A _____ _____ is a number whose square root is a natural number.
   FPTREEC   RQAUES

3. The symbol $\sqrt{}$ indicates the _____ _____ of a number.
   UEQASR   OTRO

Example 1: Evaluate Square Roots

Find the side length of a square with the given area.

a) 100 cm²   b) 62.41 cm²

Solution

a) 10 × 10 = 100
So, $\sqrt{100}$ = 10
The side length is 10 cm.

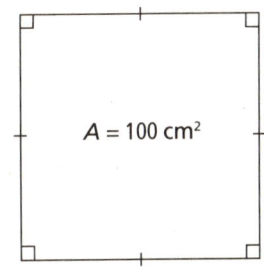

b) 7 × 7 = 49   Too low.
8 × 8 = 64   Too high.
So, $\sqrt{62.41}$ is between 7 and 8.
Use a calculator. [C] 62.41 [√x] 7.9
The side length is 7.9 cm.

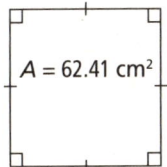

Example 2: Find Perfect Squares

Decide if each number is a perfect square.

a) 169   b) 55

Solution

a) Try the number 12.
12 × 12 = 144   Too low.
Try 13.
13 × 13 = 169   Correct!
So, $\sqrt{169}$ = 13
The square root of 169 is a natural number, 13.
So, 169 is a perfect square.

b) Try the natural number 7.

$7 \times 7 = 49$ Too low.

Try 8.

$8 \times 8 = 64$ Too high.

Since 55 is between 49 and 64, the square root of 55 is between 7 and 8.

The square root of 55 is not a natural number.

So, 55 is not a perfect square.

Check using a calculator.

[c] 55 [√x] 7.416198487

Practise

1. Evaluate the following. Use your calculator if necessary.

   a) $\sqrt{81} =$ b) $\sqrt{441} =$ c) $\sqrt{4} =$

   d) $\sqrt{1.69} =$ e) $\sqrt{30.25} =$ f) $\sqrt{19.36} =$

Hint

Key sequences may vary.

On some calculators, you need to enter [c] 100 [√] to find $\sqrt{100}$.

On other calculators, you need to enter [c] [√] 100 [=].

2. Colour each area according to the description of its square root.

   a) Square root between 5 and 6 — RED.

   b) Square root between 3 and 5 — YELLOW

   c) Perfect square — ORANGE

   d) Square root greater than 6 — BLUE

   e) Square root less than 3 — BLACK

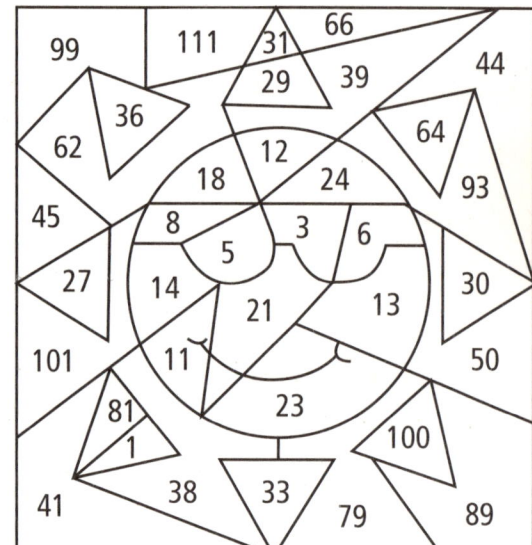

Apply

3. The Robinsons' square backyard has an area of 484 m². There is a fence around the backyard. Their square pool has an area of 144 m². The pool is in the centre of the backyard. How far is the edge of the pool from the fence? Show your work.

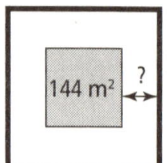

Distance =

7.3 Understand the Use of Exponents

Student Text pp. 218–223

Key Ideas Review

Match the expressions with the numbers to fill in the blanks and label the diagram.

> base (7^3)   exponent (3^5)   multiplication (8^3)   power (2^4)

- Repeated _____ (512) can be represented using exponents.

For example:

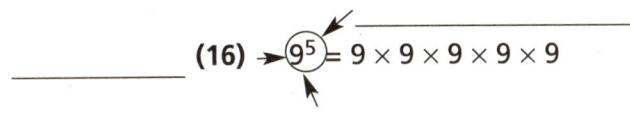

_____ (243)

_____ (16) → 9^5 = 9 × 9 × 9 × 9 × 9

_____ (343)

Example: Write and Evaluate Powers

Write each expression as a power. Then evaluate it.

a) 5 × 5 × 5 × 5

b) 4 × 4 × 4 × 4 × 4 × 4

Solution

a) 5 × 5 × 5 × 5 = 5^4
The power is five to the fourth.

Method 1: Use Paper and Pencil

5^4 = 5 × 5 × 5 × 5
  = 25 × 5 × 5
  = 125 × 5
  = 625

Method 2: Use a Calculator

[C] 5 [y^x] 4 [=] 625

b) 4 × 4 × 4 × 4 × 4 × 4 = 4^6
The power is four to the sixth.

Method 1: Use Paper and Pencil

4^6 = 4 × 4 × 4 × 4 × 4 × 4
  = 16 × 16 × 16
  = 256 × 16
  = 4096

Method 2: Use a Calculator

[C] 4 [y^x] 6 [=] 4096

Name: _____ Date: _____

Practise

1. Complete the chart. Use your calculator where necessary.

Repeated Multiplication	Exponential Form	Standard Form
	4^3	
	3^6	
$12 \times 12 \times 12$		
	8^5	
$5 \times 5 \times 5 \times 5$		
		49

2. Match the items in Column A with those in Column B.

A	B
$2 \times 2 \times 2 \times 2$	3^5
81	4^4
$3 \times 3 \times 3 \times 3 \times 3$	16
5^4	9×9
256	625

Hint

On some calculators, the power key will appear as y^x.

On others, the power key may appear as a^b or x^y.

Apply

3. Evaluate each expression to find the answer to the riddle.

What did the oak say to the maple when he was accused of lying?

8^4	$\sqrt{81}$	2.73^3	$4 \times 4 \times 4 \times 4$	6^3	$\sqrt{64}$	$10 \times 10 \times 10$
L	V	R	I	E	A	B

"Don't worry – I ___ ___ - ___ ___ ___ ___ ___ you!"
    1000 216   4096 216   8   9   216

4. Find the unknown number in each question.

a) $2^4 =$ _____

b) $3^{\underline{}} = 81$

c) $\underline{}^2 = 121$

d) $4^{\underline{}} = 1024$

7.4 Fermi Problems

Student Text pp. 224–227

Key Ideas Review

Each word in the list equals an expression. Evaluate the expressions.
Match the expressions with the numbers to fill in the blanks.

> missing = $\sqrt{169}$ large = 23   simple = 112   several = $9 \times 9 \times 9$
> decimal = $3 \times 3 \times 3$ exponent = $\sqrt{324}$ estimation = $\sqrt{81}$ reasonable = 262

1. Fermi problems are _____ (9) problems that may involve _____ (8) numbers.

2. To solve Fermi problems, you need to research _____ (13) information and make _____ (676) assumptions.

3. A Fermi problem can have _____ (729) solutions.

Example: Basketballs

How many basketballs can fit inside Junita's bedroom?

Solution

Method 1: Use the Volumes

Plan: Find the volume of a basketball.
        Find the volume of Junita's bedroom.
        Divide the volume of a basketball into the volume of Junita's bedroom.

Research: Find the diameter of a basketball. (Measure or use Internet.) *Diameter ≐ 24.3 cm*
        Assume that the volume of the ball is about the same as the volume of a cube with a side length of 24.3 cm.
        Approximate volume of one basketball = 24.3 cm × 24.3 cm × 24.3 cm
        ≐ 14 349 cm^3
        Determine the measurements of Junita's bedroom.
        It is a rectangular prism, 4 m long by 5 m wide and 2.5 m high.
        Convert to centimetres by multiplying the measurements.
        4 m = 4 × 100 cm   5 m = 5 × 100 cm   2.5 m = 2.5 × 100 cm
        = 400 cm   = 500 cm   = 250 cm
        Volume of Junita's bedroom = 400 cm × 500 cm × 250 cm
        = 50 000 000 cm^3

Calculate: Number of basketballs needed = $\dfrac{50\ 000\ 000}{14\ 349}$
        ≐ 3485

About 3485 basketballs are needed to fill Junita's room.

Method 2: Solve a Simpler Problem

Estimate the number of basketballs that would fit into a 1-m^3 cube.

The side length of a 1-m^3 cube is 1 m.
The diameter of a basketball is 24.3 cm.

Change the side length of the cube to centimetres.   1 m = 100 cm

Number of basketballs that will fit along each side of the cube = 100 ÷ 24.3
$\doteq$ 4

Find the total number of basketballs inside the cube.   4 × 4 × 4 = 64

Volume of Junita's bedroom = 4 m × 5 m × 2.5 m
= 50 m^3

This volume is 50 times the volume of the 1-m^3 cube.

Number of basketballs needed to fill Junita's bedroom = 50 × 64
= 3200

Although the two answers are different, they differ only by 285 and both round to 3000, so both answers seem reasonable.

Apply

In each question, explain and justify the process that you used to get the result.

1. How many CDs (laid flat) would it take to cover the entire area of the Yukon?

    Number of CDs =

2. How many soccer balls would it take to fill the entire volume of Skydome in Toronto?

    Number of soccer balls =

    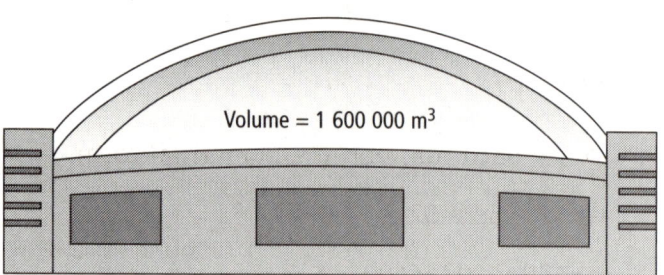

    Volume = 1 600 000 m^3

Hint

If you can't research exact information, make reasonable estimates.

Chapter 7: Reviewing for the Test

7.1 Understand Exponents
page 210

1. Find the area of the square.

   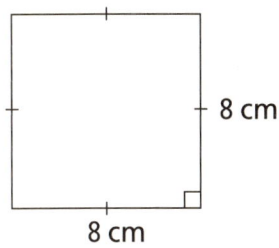

   Area =

2. Find the volume of the cube.

   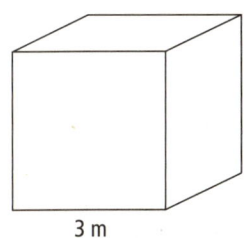

   Volume =

3. Use a calculator to evaluate.

   a) $15^2 =$ b) $5.4^2 =$

   c) $2.7^3 =$ d) $9^3 =$

4. Write each number in exponential form.

   a) $62 \times 62 \times 62 \times 62 \times 62 \times 62 =$

   b) $107 \times 107 \times 107 \times 107 =$

5. Fill in each blank with >, <, or = .

   a) 2^3 ___ 3^2 b) 5^2 ___ 3^3

   c) 1^3 ___ 1^5 d) 10^1 ___ 2^2

7.2 Represent and Evaluate Square Roots
page 214

6. List all perfect squares between 50 and 150. Explain your reasoning.

7. Evaluate.

   a) $\sqrt{49} =$ b) $\sqrt{100} =$

   c) $\sqrt{59.29} =$ d) $\sqrt{169} =$

8.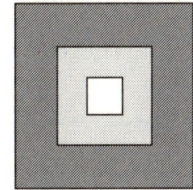

   The area of the middle square is four times the area of the smallest square. The area of the largest square is nine times the area of the middle square. The area of the largest square is 144 cm². What is the side length of the smallest square?

   Side length of smallest square =

7.3 Understand the Use of Exponents
page 218

9. Write each as a power.

   a) 729 as a power of 9 _____

   b) 1024 as a power of 2 _____

   c) 1296 as a power of 6 _____

10. Fill in each blank with >, <, or = as appropriate.

    a) 1^{20} ___ 20^1

    b) $\sqrt{36}$ ___ 3^2

    c) 2^3 ___ $\sqrt{64}$

    d) $5 \times 5 \times 5$ ___ 11^2

11. Brianna's father gave her 3¢ for her first allowance. He said he would give her triple the amount of the previous week for each week for a year. So in the first week, Brianna received 3¢, in the second week she received 9¢, in the third week she received 27¢, and so on.

    How much did Brianna receive for her allowance in the ninth week? Explain your answer.

    Brianna's allowance =

7.4 Fermi Problems
page 224

12. How many cups of sugar cubes does it take to get to the moon? Assume that each sugar cube has an edge length of 1 cm and that they are stacked one on top of the other.

    Explain and justify the process that you used to get your answer.

    Hint
    1 cup = 236.519 mL

8.1 Explore Three-Dimensional Figures

Student Text pp. 236–241

Key Ideas Review

Use the flow chart to classify the figures shown.
A figure may have more than one label.

  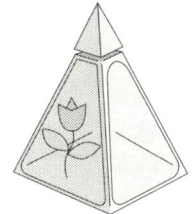 

three-dimensional figures

polyhedra

faces all polygons

prisms

base and top are parallel polygons that are congruent, other faces are rectangles

pyramids

base is a polygon, other faces are triangles that meet at one point

non-polyhedra

some faces are curved

congruent
- the same shape and size

Example 1: Classify Three-Dimensional Objects

Give the name of the figure that each object most resembles.

a)    b)    c)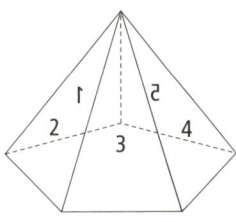

Solution

a) The safety pylon has a curved surface like this.

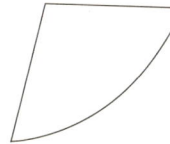

It has a base like this.

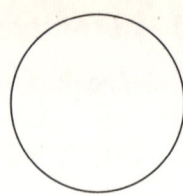

It is a cone.

b) The chocolate box has two faces like this.
This is a triangle.

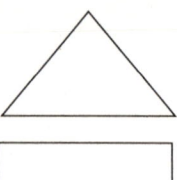

There are three faces like this.
This is a rectangle

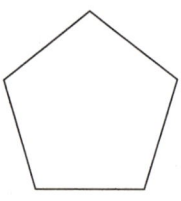

It is a triangular prism.

c) The game piece has a base like this.
This is a pentagon.

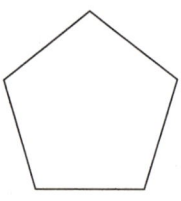

There are five faces like this.
This is a triangle.

It is a pentagonal pyramid.

Example 2: Properties of Polyhedra

The Toble Choc box is a triangular prism.

a) Name three edges of equal length.

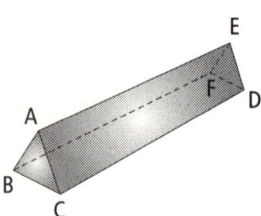

b) Name one pair of congruent faces.

edge
• where two faces meet

vertex
• where two or more edges meet

face
• flat or curved surface of object

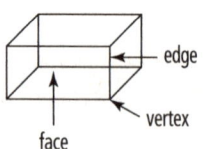

Solution

a) AE = CD = BF

b) Faces ABC and EFD are congruent.

76 MHR • Chapter 8: Three-Dimensional Geometry and Measurement

Name: _____ Date: _____

Practise

1. Identify the three-dimensional objects that make up Rover's:

   head   ears

   upper body   lower body

   arms   legs

   feet

2. a) Name the polygons that make up Rover's lower body.

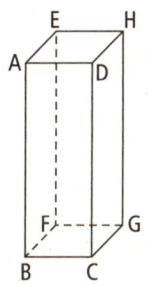

   b) Name four edges of congruent length in Rover's upper body.

   c) Name four pairs of congruent faces in Rover's upper body.

Study Skills

Create a three-dimensional figure. Exchange figures with a study partner. Name congruent edges and faces

Making Connections

Look for three-dimensional figures in your home and outside.

Make a list of the things that you see.

For example:
• a can of juice is a cylinder
• a cereal box is a rectangular prism
• a sugar cube is a cube.

Using things that you know will help you identify three-dimensional figures.

8.2 Sketch Front, Top, and Side Views

Student Text pp. 242–246

Key Ideas Review

1. Drawings of the front view, top view, and side view of a three-dimensional figure show how the figure appears from these viewpoints.

   The views are two-dimensional drawings.

   Label the front, top, and side of the robot drawing.

Example 1: Draw Front, Top, and Side Views

Draw the front, top, and side views of the game key. The front side is shaded.

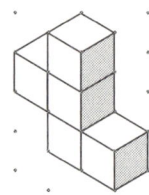

Solution

Use sugar cubes and tape to make a model of the key.
Label the front, top, and side views of the key.
Then draw the views on grid paper.

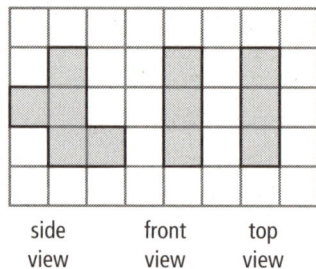

side view   front view   top view

Example 2: Draw Views of a More Complex Shape

Draw the front view, top view, and side view of the object.

Solution

Use linking cubes to make a model of the object.
Label the front, top, and side views of the object.
Then draw the views on grid paper.

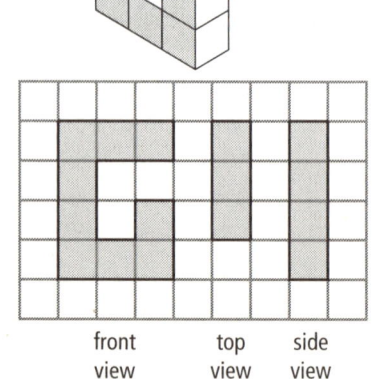

front view   top view   side view

Name: _____ Date: _____

Practise

1. Draw the front view, top view, and side view for each game piece.

a)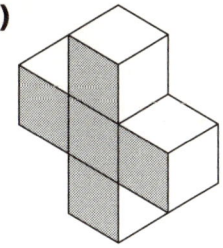

front view   top view   side view

b)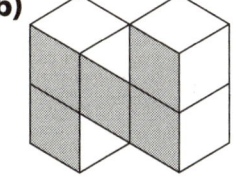

front view   top view   side view

Apply

2. The front, top, and side views of an object are shown. Sketch the three-dimensional object.

top view   front view   side view

Hint

All vertical lines should be parallel and equally spaced.

All horizontal lines should be parallel and equally spaced.

Join the dots to create lines.

3. Draw the front, top, and side views of a mug. Be sure to include any decorations.

Study Skills

Draw the front, top, and side views of several objects.

Exchange drawings with a study partner.

Draw or describe the shape of your partner's objects.

8.2 Sketch Front, Top, and Side Views • MHR 79

Name: _____ Date: _____

8.3 Draw and Construct Three-Dimensional Figures Using Nets

Student Text pp. 247–251

Key Ideas Review

Draw a line to connect each net with its polyhedron.

Net   **Polyhedron**

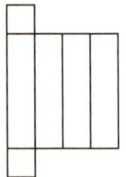   

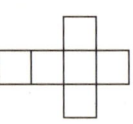

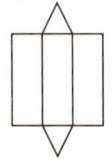   

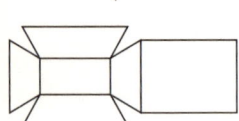   

Example 1: Drawing Nets

Draw two possible nets for the snack box.

Label the faces and measurements on your net.

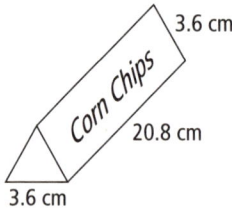

3.6 cm
Corn Chips
20.8 cm
3.6 cm

Solution

a) 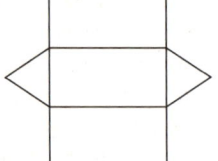   b)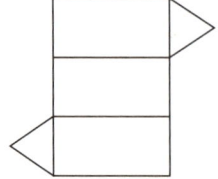

80   MHR • Chapter 8: Three-Dimensional Geometry and Measurement

Name: _____ Date: _____

Example 2: Sketching Three-Dimensional Objects From Nets

Sketch the three-dimensional object that can be made from this net.

What type of figure is formed?

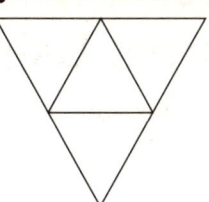

Solution

Method 1: Trace the Net and Build the Object

Trace the net onto paper.

Cut out the net. Fold each face upward along the side of the base. Tape the edges of the faces together.

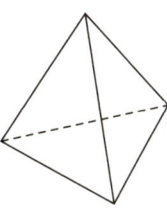

Method 2: Visualize the Object

The net has an equilateral triangle for the base and three congruent triangle faces.

When the triangles are folded up, a triangular pyramid is formed.

Practise

1. Draw the net for this object.

   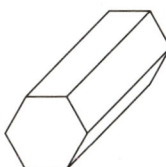

   Name the polygons that make up the sides and ends. _____

2. Sketch the three-dimensional figure for this net.

Apply

3. Which is the net for a cube? Check by tracing, cutting out, and folding the nets. Tape the edges.

   a) 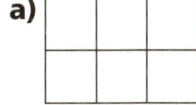   b) 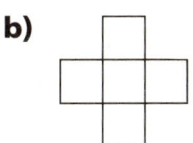   c)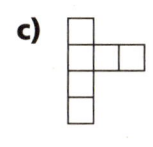

Making Connections

Find a box that is a cube. Cut open the box for its net. Compare this net with the ones shown. Which net is correct? Use other three-dimensional figures that you can find at home. What are their nets?

8.4 Surface Area of a Rectangular Prism
Student Text pp 252–257

Key Ideas Review

Fill in the blanks with the words and expressions from the list.

> faces   S.A. = 2(l × w) + 2(l × h) + 2(w × h)   sum   l × w   l × h   w × h

1. The surface area of a prism is the _____ of the areas of all its _____.

2. Complete the area formulas for the faces.

3. Surface area of a rectangular prism =
   _____

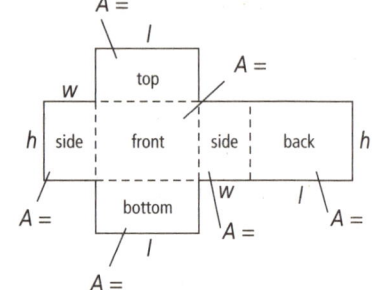

Example 1: Find Surface Area

Find the surface area of the gift box.

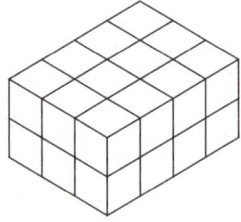

Solution

Method 1: Count the Number of Faces on Each View

front and back views: 8 + 8 = 16

top and bottom views: 12 + 12 = 24

side views: 6 + 6 = 12

Total number of faces: 16 + 24 + 12 = 52

Each face has an area of 1 cm².

The surface area is 52 cm².

Check by building a model out of sugar cubes.

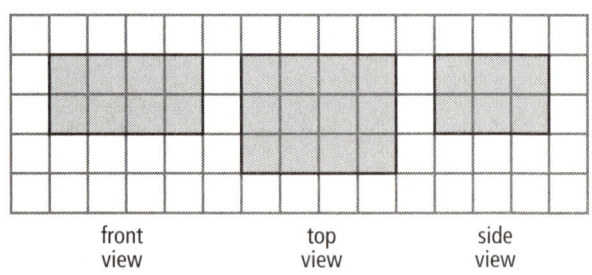

Method 2: Calculate the Surface Area

Front face: Top face: Left-side face:

l = 2   l = 3   l = 2
w = 4   l = 4   l = 3

82   MHR • Chapter 8: Three-Dimensional Geometry and Measurement

Name: _____ Date: _____

Front face:
$A = l \times w$
$A = 2 \times 4$
$A = 8$
The back face has the same area as the front face, 8 cm².

Top face:
$A = l \times w$
$A = 3 \times 4$
$A = 12$
The base has the same area as the top face, 12 cm².

Left-side face:
$A = l \times w$
$A = 2 \times 3$
$A = 6$
The right-side face has the same area as the left-side face, 6 cm².

Surface area = front + back + top + base + left side + right side
 S.A. = 8 + 8 + 12 + 12 + 6 + 6
 S.A. = 52

The surface area is 52 cm².

Method 3: Use a Formula

$l = 3$, $w = 4$, and $h = 2$.

S.A. = $2(l \times w) + 2(l \times h) + 2(w \times h)$
S.A. = $2(3 \times 4) + 2(3 \times 2) + 2(4 \times 2)$
S.A. = 24 + 12 + 16
S.A. = 52

The surface area is 52 cm².

Practise

1. Calculate the surface area of each rectangular prism.

a)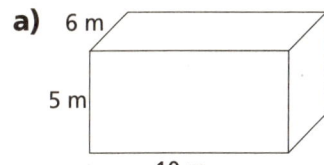

b)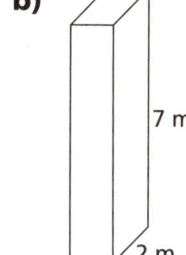

c)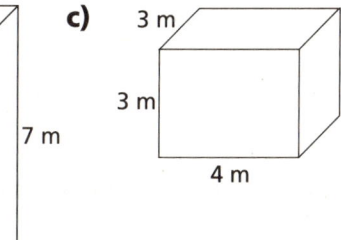

a) S.A. = _____ m²   b) S.A. = _____ m²   c) S.A. = _____ m²

Apply

2. The sides and ceiling of the greenhouse are made of glass. How much surface area is covered by glass?

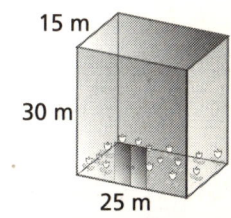

3. The length of a recreation room is 5 m, the height is 3 m, and the width is 7 m. The surface area is 106 m² when counting the number of 1-m² squares and 107 m² when using the surface area formula. Which surface area is correct? Explain.

Hint
Do not include the area of the floor.

8.5 Volume of a Rectangular Prism

Student Text pp 258–261

Key Ideas Review

Fill in the blanks with the words or expressions from the list.

> height   area of base × height   space

1. Volume is the amount of _____ occupied by an object.

2. The volume of a rectangular prism can be found by length × width × _____.

3. The volume of a rectangular prism can also be found by _____.

Example 1: Calculate Volume

Find the volume of the juice box.

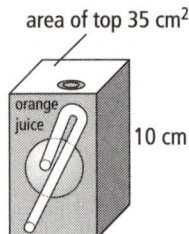

Solution

Volume = area of base × height
$V = 35 \times 10$
$V = 350$

The volume of the juice box is 350 cm³.

Hint
area of base = area of top

Example 2: Solve a Problem Involving Volume

Find the volume of three juice boxes.

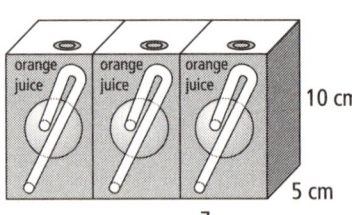

Solution

Find the volume of one juice box.

$l = 5$ cm, $w = 7$ cm, $h = 10$ cm

Volume = area of base × height
Volume = length × width × height
$V = 5 \times 7 \times 10$
$V = 350$

The volume of one juice box is 350 cm³.

The volume of three juice boxes is 3 × 350 cm³, or 1050 cm³.

84   MHR • Chapter 8: Three-Dimensional Geometry and Measurement

Name: _____ Date: _____

Practise

1. Calculate the volume of each rectangular prism.

a)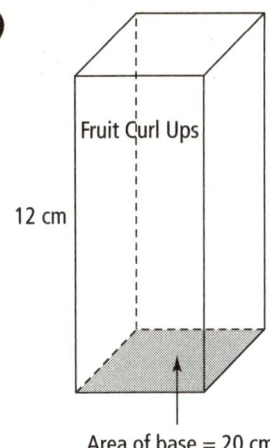

12 cm

Area of base = 20 cm²

V = _____ cm³

b)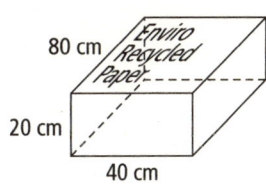

80 cm, 20 cm, 40 cm

V = _____ cm³

c)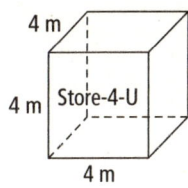

4 m, 4 m, 4 m

V = _____ m³

Apply

2. James buys a toy car that will fit into the box. He wants to buy four cars to send to his friend. What is the volume of the box that he will need?

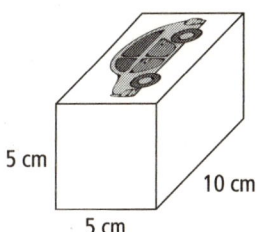

5 cm, 5 cm, 10 cm

3. Rose drank one quarter of the apple juice in a large juice box. The box measures 20 cm by 12 cm by 8 cm. What volume of juice is left?

Chapter 8: Reviewing for the Test

8.1 Explore Three Dimensional Figures
page 236

1. Name the three-dimensional figure that matches each description.

   a) a die

   b) one pentagonal face and five triangular faces

   c) a wedge of cheese

   d) a juice can

8.2 Sketch Front, Top, and Side Views
page 242

2. a) Create the letter C out of five sugar cubes. Make a sketch of your model.

   b) Draw the front view, top view, and side view of your model.

   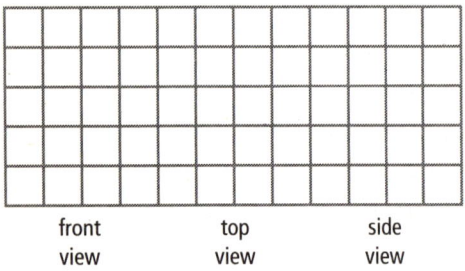

   front view   top view   side view

8.3 Draw and Construct Three-Dimensional Figures Using Nets
page 247

3. Label the top, bottom, front, back, left side, and right side of the block. The front has a star.

   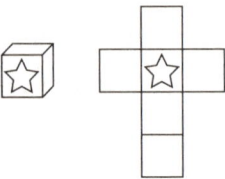

8.4 Surface Area of a Rectangular Prism
page 252

4. How much wrapping paper is needed to cover the box?

   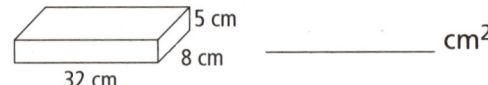

   _____ cm²

8.5 Volume of a Rectangular Prism
page 258

5. Which vase will hold the most water?

   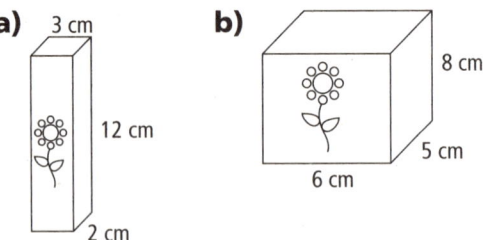

6. Apple Juice Co. has 100 000 cm³ of juice. Each of its juice boxes is 5 cm by 5 cm by 10 cm. How many juice boxes can be filled?

86 MHR • Chapter 8: Three-Dimensional Geometry and Measurement

9.1 Collect and Organize Data

Student Text pp. 274–279

Key Ideas Review

Match the terms in Column A with the definitions in Column B.

A	B
1. primary data	• compares the frequencies of different parts of a data set
2. pictograph	• useful for comparing data
3. bar graph	• visually appealing, but may not represent the data precisely
4. secondary data	• data obtained from other sources
5. tally charts and frequency tables	• data collected by surveying or counting

Example 1: Use a Frequency Table to Draw a Bar Graph and a Pictograph

One brand of jelly beans comes in seven colours. The number of each colour of jelly beans in four boxes was found. The results are shown.

Colour	Tally
Red	++++ ++++ ++++ ++++ ++++ ++++
Blue	++++ ++++ ++++ IIII
Orange	++++ ++++ ++++ ++++ ++++ ++++ III
Yellow	++++ ++++ ++++ ++++ ++++ II
Pink	++++ ++++ ++++ ++++ ++++ ++++ I
Purple	++++ ++++ ++++ ++++ II
Green	++++ ++++ ++++ ++++ ++++ ++++ I

a) Draw a bar graph to show the data.

b) Draw a pictograph to show the data.

Solution

a) Organize the data in a frequency table.

Colour	Tally	Frequency
Red	++++ ++++ ++++ ++++ ++++ ++++	30
Blue	++++ ++++ ++++ IIII	19
Orange	++++ ++++ ++++ ++++ ++++ ++++ III	33
Yellow	++++ ++++ ++++ ++++ ++++ II	27
Pink	++++ ++++ ++++ ++++ ++++ ++++ I	31
Purple	++++ ++++ ++++ ++++ II	22
Green	++++ ++++ ++++ ++++ ++++ ++++ I	31

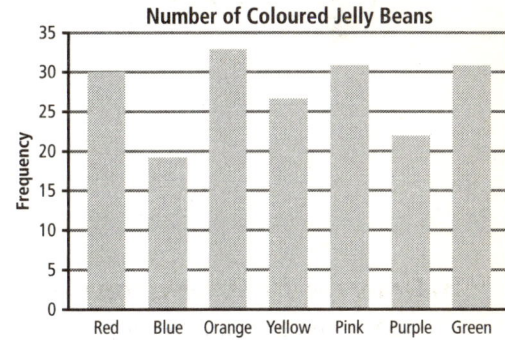

b) Use a jelly bean symbol.

Let one 🫘 represent five beans.

Divide by 5 to find the number of 🫘 symbols to use for each value.

Number of Coloured Jelly Beans

Red	🫘🫘🫘🫘🫘🫘
Blue	🫘🫘🫘🫘
Orange	🫘🫘🫘🫘🫘🫘🫘
Yellow	🫘🫘🫘🫘🫘
Pink	🫘🫘🫘🫘🫘🫘
Purple	🫘🫘🫘🫘
Green	🫘🫘🫘🫘🫘🫘

🫘 represents five jelly beans

Name: _____ Date: _____

Practise

1. Some students were surveyed about their favourite season.

   a) Display the data using a bar graph.

Season	Tally	Frequency
Winter	|||| |||	8
Spring	||||		6
Summer	|||| |||| ||||	14
Fall	||||	4

   b) Which is the most popular season? _____

   c) How many students were surveyed? _____

2. A greeting card company conducted a survey to discover which holiday is the most popular among students.

   a) Complete the frequency table.

   b) Display the data in a pictograph.

Season	Tally	Frequency
Easter	|||| |||	
Christmas	|||| |||| |||| |||| ||	
Your Birthday	|||| |||| ||||		
Halloween	|||| ||	
Other	||||	

   c) Explain how you decided how many people each symbol should represent.

3. Classify each of the following as primary data or secondary data.

   a) Mark records the number of shots on goal during a hockey game.

   b) Samantha researches the number of grizzly bears living in North America.

   c) Juan learns from the travel agency the most popular places to travel for summer holidays.

   d) Nalina counts the number of cars that go through an intersection between 5 p.m. and 6 p.m.

9.2 Stem-and-Leaf Plots

Student Text pp. 280–285

Key Ideas Review

Fill in the blanks with words from the list.

> leaves   Stem   stem-and-leaf plot   increasing   Leaf   stems

1. A _____ is used to organize and order large sets of numeric data.

2.
   a) _____ (tens)   b) _____ (ones)

Stem	Leaf
0	3 7
1	4 6 9
2	0 0 2 4 8
3	7 7 7
4	4 8 9 9
5	1 3 3 4

   The data is organized into groups called c) _____.

   The d) _____ in each stem are written in e) _____ order.

Example 1: Read and Interpret a Stem-and-Leaf Plot

The following stem-and-leaf plot shows the high temperatures in degrees Celsius for two weeks in May in Southern Ontario.

Stem (tens)	Leaf (ones)
0	8 9 9
1	0 1 1 2 5 7 8 9
2	2 3 4

a) How many days was the temperature in the 20s? What were these temperatures?

b) Which stem contains the most data?

c) How many days was the temperature below 16°C?

Solution

a)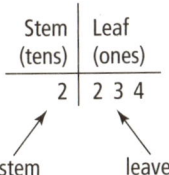

Stem (tens) | Leaf (ones): 2 | 2 3 4
↑ stem   ↑ leaves

Look in the stem column for 2. There are three leaves in the stem for 2. So there were three days when the temperature was in the 20s. The temperatures were 22°C, 23°C, and 24°C.

b) Stem 1 contains the most data. It has eight leaves.

c) Colder than 16°C means all of stem 0 and the part of stem 1 that has leaves less than 6. Stem 0 has three leaves. Five of the leaves in stem 1 are less than 6. There were 3 + 5 = 8 days where the temperature was less than 16°C.

Name: _____ Date: _____

Example 2: Create a Stem-and-Leaf Plot

The number of points that the Twirlin' Tornadoes were able to score in each of their football games for the regular season is shown below.

Create a stem-and-leaf plot to show the data.

24	14	22	32	19	7	31	27	27	18
28	18	24	23	29	31	15	24	9	37

Solution

Use the tens digit for the stem.
Use the ones digit for the leaves.
First organize the numbers by their stems.

Stem (tens)	Leaf (ones)
0	7 9
1	4 9 8 8 5
2	4 2 7 7 8 4 3 9 4
3	2 1 1 7

Now rearrange the leaves in each stem to show them in increasing order.

Stem (tens)	Leaf (ones)
0	7 9
1	4 5 8 8 9
2	2 3 4 4 4 7 7 8 9
3	1 1 2 7

Practise

1. Use the stem-and-leaf plot from Example 2 to answer the questions.

   a) How many leaves does stem 2 have?

   b) What scores are represented by stem 0?

   c) What was the highest score?

   d) How many times did the Twirlin' Tornadoes score more than 25 points?

Apply

2. Recent scores on a video game by a group of friends are shown.

55	67	89	99	84	67	72	93	77	84	84
72	96	83	72	55	66	96	59	89	80	

   a) Create a stem-and-leaf plot for this data.

   b) What were the most frequent scores?

   c) Players who score over 85 points get to move on to a harder level. How many players in this group got to play the harder level?

90   MHR • Chapter 9: Data Management: Collection and Display

Name: _____ Date: _____

9.3 Circle Graphs

Student Text pp. 286–291

Example 1: Draw a Circle Graph

40 students were surveyed to find their favourite activity at the beach.

Draw a circle graph to display the data.

Beach Activity	Number of People
Building sandcastles	10
Playing sports	3
Swimming	20
Tanning	7

Solution

Method 1: Work With the Fractions You Know

Beach Activity	Number of People	Fraction	Size of Section
Swimming	20	$\frac{20}{40}$	This is the same as one-half.
Building sandcastles	10	$\frac{10}{40}$	This is the same as one-quarter.

Method 2: Calculate Section Angles

Beach Activity	Number of People	Fraction	Decimal	Section Angle
Tanning	7	$\frac{7}{40}$	$7 \div 40 = 0.175$	$0.175 \times 360° = 63°$
Playing sports	3	$\frac{3}{40}$	$3 \div 40 = 0.075$	$0.075 \times 360° = 27°$

Draw a circle. Use a protractor to measure each section.

Shade or colour the sections if you wish.

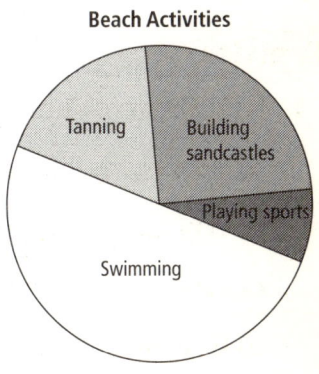
Beach Activities

Example 2: Read and Interpret a Circle Graph

Kanmeer made a circle graph of how he spent his time one day.

How many hours did Kanmeer spend at school or doing homework that day?

Solution

Kanmeer spent 35% of his day at school or doing homework.

There are 24 hours in a day.

Find 35% of 24.

35% of 24 = 0.35×24
           = 1.8 hours

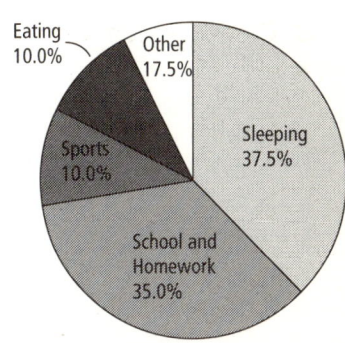
Kanmeer's Daily Activities

Practise

1. 60 people were surveyed about their favourite type of snack food.

Snack Food	Number of People	Fraction	Decimal	Section Angle
Chocolate	25			
Peanuts	10			
Chips	15			
Other				
Total	60			

   Hint
   Do not round your decimal number. Use the number on your calculator to find the section angle.

   a) Complete the table of calculations.

   b) Draw a circle graph to show the data.

2.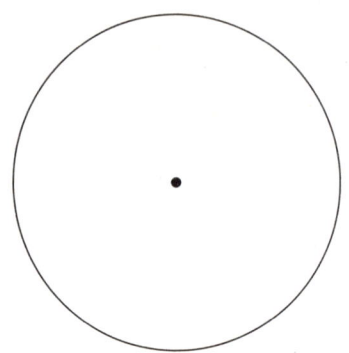

   a) Of what group of clothing does Mona have the most items in her closet?

   b) How many shorts and T-shirts does Mona have in her closet?

   c) How many dresses are in Mona's closet?

Apply

3. Jeffery is a goaltender. He asked his sister Zoe to keep some statistics for him.

   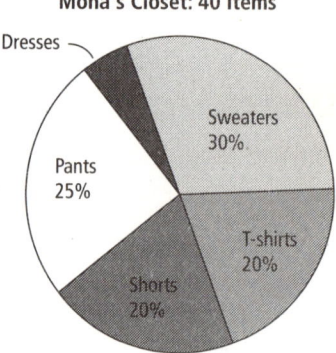

Result of Shots at Jeffery	Tally
Goals	IIII
Saves	HHT HHT HHT I
Shots blocked	HHT III
Shots missed the net	HHT HHT II

   a) Use the information in Zoe's tally sheet to create a circle graph of the data.

   b) How can Jeffery use this data to convince his coach that the team's defensive players are doing a good job? Explain your answer.

   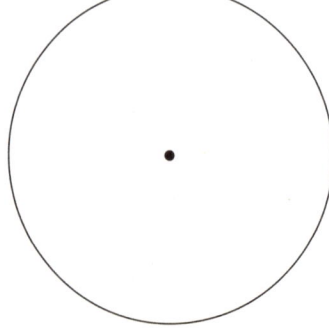

92   MHR • Chapter 9: Data Management: Collection and Display

9.4 / 9.5 Use Databases to Find Data / Use a Spreadsheet to Display Data

Student Text pp. 292–303

Key Ideas Review

1. Cross out the entry that does not apply to spreadsheets.

A type of database that allows you to select specific information or sort information	Can only store information on up to five items
A software tool used to organize and display numeric data	Can be used to develop various types of graphs

2. Circle the words that make the sentences correct.

   a) A pie chart is another name for a **bar graph / circle graph**. It shows how each part of a data set compares to the whole.

   b) A **bar graph / circle graph** shows how different parts of a data set are related.

Practise

1. Compare the two graphs from the CANSIM database.

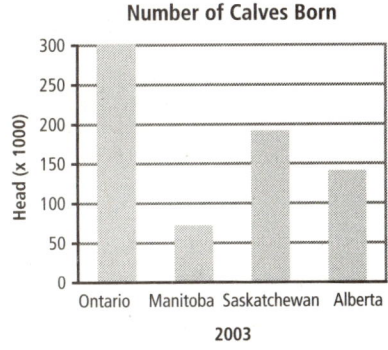
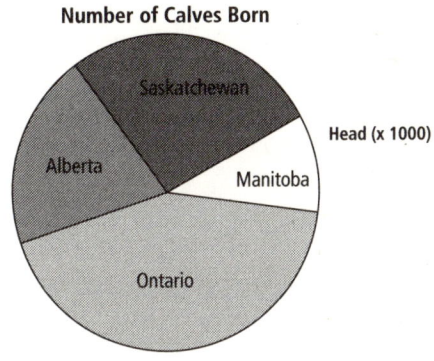

   a) Which graph would you use to determine whether there were more calves born in Ontario or in Manitoba, Alberta, and Saskatchewan combined in 2003? Explain.

   b) Which graph would you use to determine how many calves were born in Manitoba in 2003? Explain your reasoning.

2. Examine the two graphs below, which display data on how much time Cora spends on each subject for homework in a week.

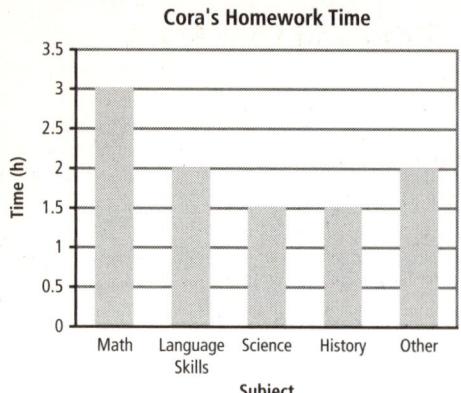
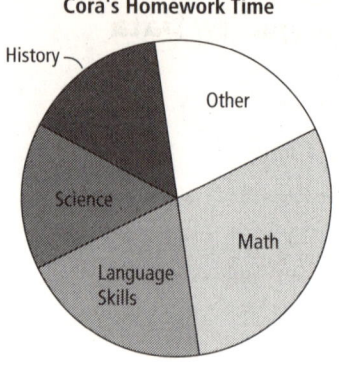

Which graph clearly shows that Cora spends most of her time doing math? Explain your answer.

Apply

3. Explain an advantage and a disadvantage for each.

   a) Using a database.

   Making Connections

   You may use a database every day and not even know it!

   A recipe box organized in alphabetical order is a database. Find other examples of databases in your home. How are they organized?

   b) Using a spreadsheet.

Chapter 9: Reviewing for the Test

9.1 Collect and Organize Data
page 274

1. Coach Rivers kept a tally of how many baskets his players made in one game.

Player	Tally	Frequency
John	|||| ||||	
Mya	|||| |||| |||	
Parker	||||	
Annelle	||||	
Robin	|||| ||||	

   a) Complete the frequency table.

   b) Draw a bar graph to display the data.

   c) Which player should be named the offensive star of the game?

   d) Is this primary or secondary data?

2. Christina ate the following number of apples in one month.

   Week 1: 6   Week 3: 7
   Week 2: 2   Week 4: 4

   a) Draw a pictograph to display the data.

   b) How many apples did Christina eat during this entire month?

9.2 Stem-and-Leaf Plots
page 280

3. The stem-and-leaf plot shows the number of trick-or-treaters each house on Maple St. received last year.

Stem (tens)	Leaf (ones)
2	6 7 8 8
3	0 3 3 3 7
4	2 8
5	3 4 5

   a) How many houses received between 30 and 39 trick-or-treaters?

   b) How many trick-or-treaters visited the house that received the most trick-or-treaters?

   c) How many houses are on Maple St.?

4. The scores resulting from throwing three darts are shown for 24 players.

28	18	57	64	20	62
22	37	43	41	55	28
35	48	34	54	32	56
41	53	27	48	61	18

   a) Create a stem-and-leaf plot for this data.

   b) How many players threw a score of 48?

   c) How many players received a score of 50 or more?

9.3 Circle Graphs
page 286

5. Bill surveyed Grade 7 classes to find out what types of television shows they most like to watch.

Type of Show	Number of People
Comedy	10
Reality	15
Sports	20
Other	5

a) Complete the circle graph to show this data.

6. Jethro made a circle graph of the types of seeds that were planted in his garden.

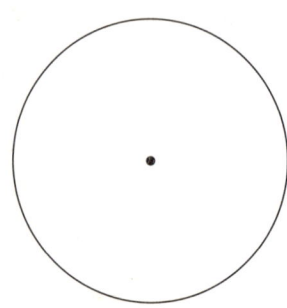
Types of Seeds
Watermelon 5%
Carrots
Lettuce 20%
Rhubarb 35%
Tomatoes 25%

a) What percent of the garden has carrot seeds?

c) Eighty seeds were planted. How many seeds of each type were used?
Carrots: ____   Lettuce: ____
Rhubarb: ____   Tomatoes: ____
Watermelon: ____

9.4 Use Databases to Find Data
9.5 Use a Spreadsheet to Display Data
page 292

7. Hope created a spreadsheet of how many chapters she read from her book last week.

	A	B	C	D
1	Monday	6		
2	Tuesday	5		
3	Wednesday	6		
4	Thursday	3		
5	Friday	4		
6				

a) Which cell indicates how many chapters Hope read on Thursday?

b) How many chapters did she read during the week?

c) On which day(s), did she read the most?

d) What type of graph would you choose to display the data? Explain your reasoning. Make a sketch to illustrate your answer.

96   MHR • Chapter 9: Data Management: Collection and Display

10.1 Analyse Data and Make Inferences

Student Text pp. 312–317

Key Ideas Review

Circle the correct word in each sentence.

1. A line graph shows **trends / predictions**.

2. A frequency table **graphs / organizes** a set of data.

3. You can make **points / predictions** from analysing a data set.

Example 1: Analyse a Trend

Analyse the sales trends for two hot dog stands over several years.

a) Describe the trend in sales for each hot dog stand.

b) Which hot dog stand had the greater sales for most of the years?

c) Predict which hot dog stand will sell more hot dogs in the next two years.

d) Give a possible reason for the difference in sales between stands A and B.

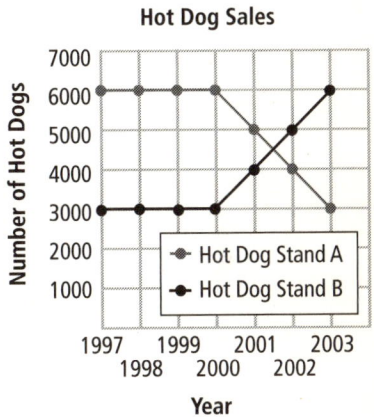

Solution

a) Hot dog stand A had a decrease in its sales over the years.
Hot dog stand B had an increase in its sales over the years.

b) Hot dog stand A sold the most hot dogs for most of the years.

c) The trend for hot dog stand B shows increasing sales over the years. Hot dog stand B will most likely sell more hot dogs in the next two years than stand A will.

d) Hot dog stand B might sell a hot dog that is more popular than stand A's. For example, stand A might sell only meat hot dogs. Stand B might sell vegetarian hot dogs, which have become more popular over the years.

Example 2: Analyse a Data Set

James catches worms to sell to a fishing store. He records the number caught over several days.

15 7 3 5 8 7 5 4 7

a) Use a frequency table to organize the data.

b) Describe the data.

Solution

a)

Number of Worms	Tally	Frequency
3	|	1
4	|	1
5	||	2
7	|||	3
8	|	1
15	|	1

b) The smallest number of worms caught is 3, and the greatest is 15. James caught 7 worms the most often.

Practise

1. Describe each trend.

   a) 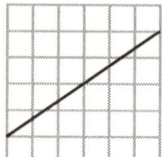   b)    c) 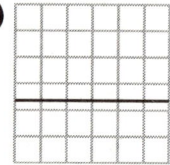   d)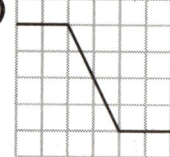

2. Anisha buys seven packages of red licorice. She counts the number of licorice sticks in each pack.

   25   27   23   25   24   27   27

   a) Use a frequency table to organize the data.

   b) Describe the data.

Apply

3. Analyse the speed trends for Jasmine and Claire in the 100-m sprint event.

   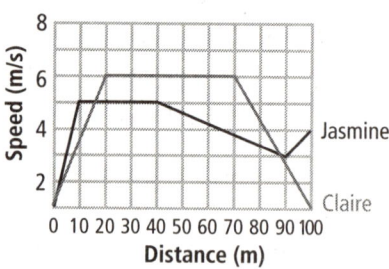

   a) Who reached her top speed first?

   b) Who really slowed down near the end of the race? Explain.

   c) When were they running at the same speed? _____ and _____

   d) Who won the race? Give reasons.

98   MHR • Chapter 10: Data Management: Analysis and Evaluation

Name: _____   Date: _____

10.2 Measures of Central Tendency

Student Text pp. 318–325

Key Ideas Review

Match each item in Column A with two items from Column B. The two items from B will be a definition and a method for finding the item in A.

Column A
1. mean
2. median
3. mode

Column B
- average
- add all values and divide by the number of values
- most common value
- order the values and find the middle one
- find the value that repeats the most
- middle value

Example: Understand Median, Mode, and Mean

Chaz is training for a long distance race. He records the number of laps he runs each day.

25   33   27   25   28   28   31   30   28   15   27

a) Find the median. Explain what it means.

b) Find the mode. Explain what it means.

c) Find the mean. Explain what it means.

Solution

a) Rearrange the data from least to greatest.
15, 25, 25, 27, 27, **28**, 28, 28, 30, 31, 33
The median is 28. This is the middle value.
You can also use a stem-and-leaf plot to find the median.
The stem shows the tens digits.
The leaves show the ones digits.

Stem (tens)	Leaf (ones)
1	5
2	5 5 7 7 8 8 8
3	0 1 3

b) The mode is the number that occurs most often. The mode is 28.
On the stem-and-leaf plot, the leaf 8 appears the most. So the mode is 28.
There were three days when Chaz ran 28 laps.

c) $\text{mean} = \dfrac{\text{sum of all values}}{\text{number of values}}$

$= \dfrac{25 + 33 + 27 + 25 + 28 + 28 + 31 + 30 + 28 + 15 + 27}{11}$

$= 27$

The mean is 27. Chaz ran 27 laps, on average, each day.

Name: _____   Date: _____

Practise

1. Find the mean, median, and mode.

a) 12, 4, 7, 13, 33, 0

Mean =

Median =

Mode =

b)

Stem (tens)	Leaf (ones)
0	3 7
1	1 7 9
2	2

Mean =

Median =

Mode =

2. a) Create a stem-and-leaf plot for the data.

    30   54   37   72
    71   72   38   54   72

Stem (tens)	Leaf (ones)

b) Find the median, mode, and mean.

median =   mode =   mean =

Apply

3. The final marks for an English class are shown.

65	70	72	80	85	88	90
77	70	80	80	70	70	71
66	65	67	65	80	70	92
71	66	68	72	88	90	70

a) Find the median, mode, and mean.

median =   mode =   mean =

b) Which measure of central tendency best describes the marks? Explain.

Name: _____   Date: _____

10.3 Bias

Student text pp. 326–330

Key Ideas Review

For each question, circle True or False.

1. Survey questions should have bias. **True   False**

2. How a question is asked can influence responses. **True   False**

3. The wording of a question can influence certain responses more than others. **True   False**

Example: Bias in Survey Questions

Read the following survey questions. Do they contain bias? If so, reword the questions to avoid bias.

a) What is your favourite subject?
A mathematics
B science
C English
D other _____

b) Should the store have more suitable hours by opening 6 days a week instead of 5 days a week?

YES   NO

c) Do you think the number of hours the public library is open is too many, about right, or too few?

Solution

a) More people might select one of the first three options because this is easiest. To remove bias, ask "What is your favourite subject?" and do not include the choices.

b) The question implies that the store has hours of operation that are not suitable. People might answer yes. To remove the bias, ask "Should the store open 6 days a week?"

c) The choices seem to be fair. This question does not seem to contain bias.

Practise

Questions 1 to 3 are biased towards a particular response. Rewrite each question to remove the bias.

1. Since the Montreal Canadiens have won 24 Stanley Cup championships, are they the best team?

YES   NO

2. Do you agree or disagree that fighting in basketball should be heavily fined?

3. What is the most popular flavour of potato chips?
  A Regular
  B Salt and vinegar
  C Barbecue
  D Other

Study Skills

Collect four or five examples of biased statements from a newspaper. Indicate the bias in each statement. Rewrite each statement so that it is unbiased. Exchange examples with a study partner. Compare your unbiased statements.

Apply

4. Football has more rules than soccer and baseball combined. Do you think this is a confusing game?
YES   NO

a) What answer do you think is expected?

b) Why is this question biased? Explain.

c) Rewrite the question so it is not biased.

10.4 Evaluate Arguments Based on Data

Student Text pp. 331–335

Key Ideas Review

Circle the correct word to complete each sentence.

1. The media often make statements that are based on **secondhand information / data**.

2. Sometimes **unimportant / important** points are intentionally left out.

3. A misleading graph is one that **accurately represents / exaggerates** a point.

Example: Misleading Graphs

The graphs show the number of T-shirts sold by two students, Ling and Karla. How does the graph exaggerate the number of T-shirts sold by Ling?

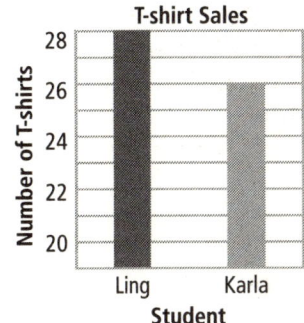

Solution

Showing just part of the vertical scale makes it appear that Ling sold about 1.5 times as many T-shirts as Karla.

The correct graph shows that Ling sold only two more T-shirts than Karla did.

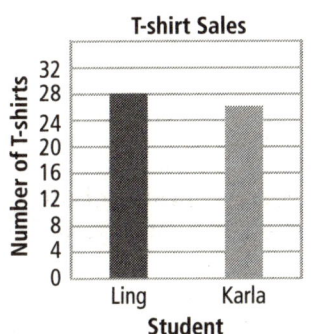

Practise

1. The graph shows the change in the price of a basketball ticket from one year to the next.

   a) What impression does the graph give?

   b) Explain why this graph is misleading.

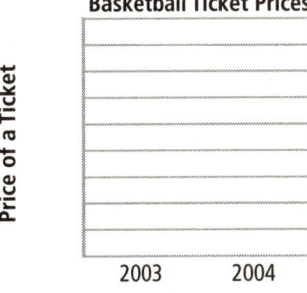

Apply

2. Draw a new graph for the data in question 1 that is not misleading.

Chapter 10: Reviewing for the Test

10.1 Analyse Data and Make Inferences
page 312

1. The graph shows the CD sales for different types of music among teens.

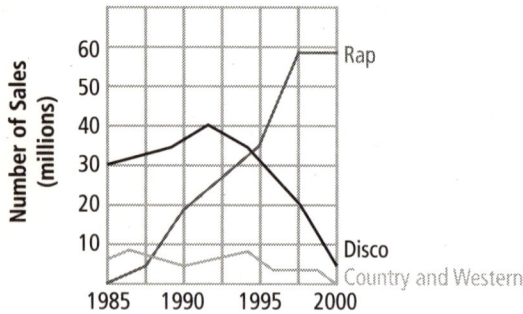

a) Describe the trend for each type of music.

b) Predict which music style will continue.

10.2 Measures of Central Tendency,
page 318

2. At a science fair, a judge gave the following scores.

11	37	46	22	45	17
37	45	13	45	28	19
45	32	48	23	17	17

a) Find the mean.

b) Use a stem-and-leaf plot to find the median and mode.

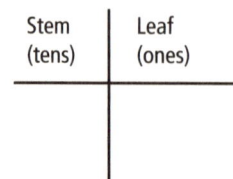

10.3 Bias
page 326

3. Your gym teacher asks, "Do you think that I've done a great job of introducing volleyball?" Explain the bias in the situation and in the question.

4. A survey asks

Which is the best Canadian hockey team?		
A Montreal	B Calgary	C Vancouver
D Ottawa	E Edmonton	F Others

   There are only six Canadian teams. Where is the bias in this question?

10.4 Evaluate Arguments Based on Data
page 331

5. Alan and Bob raised money through pledges.

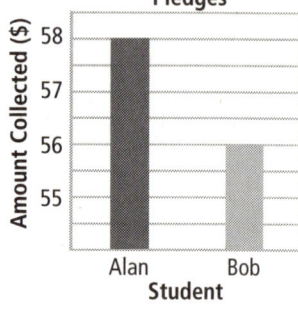

a) What impression does the graph give?

b) Draw the graph so that it is not misleading.

104   MHR • Chapter 10: Data Management: Analysis and Evaluation

11.1 Compare and Order Integers

Student Text pp. 346–351

Key Ideas Review

Fill in the blanks with the words or symbols from the list.

> less   0   opposite   >   vertical   greater   <

1. For each number line, which numbers are positive and which are negative?

   a)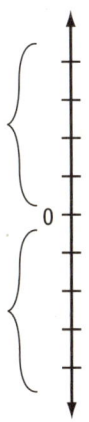

   b)

   integer
   • one of the numbers
   … –3, –2, –1, 0, +1, +2, +3 …

2. On an integer number line, _____ integers are equal distances to the left or right of _____.

3. Compare integers.

   Use the words *less* or *greater*.   Use the symbol < or >.

   –4 is _____ than –2   –4 ____ –2

   –1 is _____ than –6   –1 ____ –6

Example 1: Use a Vertical Number Line

Use a number line to show the temperatures –2, +3, +1, –1, and 0.
List the temperatures from coldest to warmest.

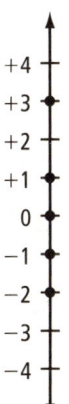

Solution

Coldest to warmest: –2, –1, 0, +1, +3

Example 2: Use a Horizontal Number Line

Compare the integers −4, +2, and +3 with the integer −1.

Use a number line, words, and symbols.

Solution

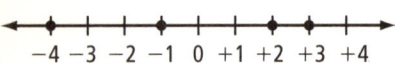

Words	Symbols
−4 is less than −1	−4 < −1
+2 is greater than −1	+2 > −1
+3 is greater than −1	+3 > −1

Literacy Connections

The symbol > is wider on the left. So a number to the left of > is larger than the number to the right. So 10 > 8 means "Ten is greater than eight."

Practise

1. **a)** Use the number line to show the integers −1, −4, +3, −2, and +1.

   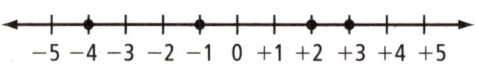

   b) Which integers are greater than 0? _____

   c) Which integers are less than 0? _____

   d) Which integers are opposite integers? _____

   e) List the integers in increasing order. _____

2. Use a < or > in each ◯ to compare the integers.

   a) +12 ◯ +11   **b)** +8 ◯ −8   **c)** −14 ◯ −13

   d) +7 ◯ 0   **e)** −3 ◯ +10   **f)** +2 ◯ −9

Apply

3. When Amy woke up on Monday morning, the temperature outside was −3°C. By noon, the temperature was +5°C.

   a) Was it warmer or colder at noon? Explain.

   b) Use the number line to find how much the temperature had changed by noon.

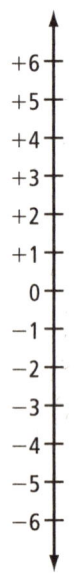

Name: _____ **Date:** _____

11.2 / 11.3 Explore Integer Addition / Adding Integers

Student Text pp. 352–361

Key Ideas Review

Circle the word to complete each sentence. Join the boxed letters in each correct word for the answer to the riddle below.

When asked if she was sure she wasn't less than zero, 7 said,

"Yes, I'm _____."

1. In the student textbook, a blue chip represents –1, and a red chip represents +1. In this workbook, ⊖ represents –1, and ⊕ represents +1. ⊖/⊕ shows the zero **rule** / **principle**.

2. To add a negative number using a vertical integer number line, move **down** / **up**.

3. The sign of the sum of ⊕ + ⊕ is **negative** / **positive**.

4. To add a positive number using a horizontal number line, move **left** / **right**.

5. To add a negative number using a horizontal number line, move **left** / **right**.

6. Coloured disks used to model integer addition are called integer **chips** / **pennies**.

7. The sign of the sum of ⊖ + ⊖ is **negative** / **positive**.

8. When adding two integers of different signs, the sum has the sign of the integer farthest from **one** / **zero**.

Example: Use a Model for Integer Addition

While playing a board game, Alyse lost four points, then gained seven points.

Method 1: Chips

lost 4 points → –4

gained 7 points → +7

A pair of opposite chips makes a zero.

Alyse has +3 points.

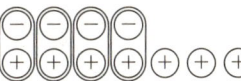

Method 2: Number Line

lost 4 points → –4

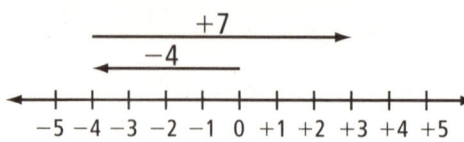

gained 7 points → +7

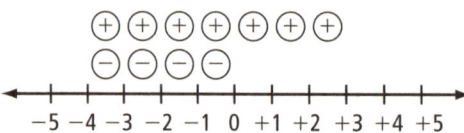

zero principle
- When pairs of opposite integers cancel each other, the result is zero.

Alyse has +3 points.

You can also show her score using integer chips.

Practise

1. What integer sum is shown? Give each result.

a)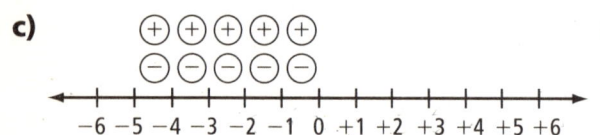

b)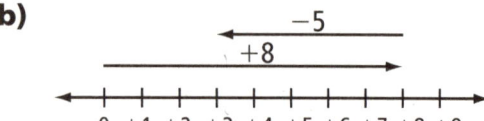

c) (image of chips and number line)

d)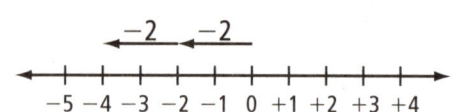

2. Use mental arithmetic to find each sum.
Check your answer using integer chips or a number line.

a) (+6) + (–8) =

b) (–7) + (–1) =

c) (0) + (–3) =

d) (+3) + (–8) =

e) (–10) + (+12) =

f) (+9) + (–15) =

108   MHR • Chapter 11: Integers

Name: _____   Date: _____

Apply

3. Shahan played a game by flipping a coin. For every head he got a point, and for every tail he took away one point.

   a) After 10 tosses, heads had come up four times. Show this value on the number line.

   b) The other six tosses were all tails. Show this value on the number line.

   c) Show Shahan's score if he then tosses three heads in a row.

   d) Write the addition statement. Then, give the result.

   (number line from +5 to −5)

4. Write an integer addition expression for each situation. Then, find the sum.

   a) At 11 a.m., Ben set his clock back one hour.

   b) A diver is 5 m above the water. She dives down 8 m from her starting position.

   c) The temperature is −3°C, goes up 12 degrees, then drops 14 degrees.

5. Describe each addition statement in money terms. Find and interpret each sum.

   a) (+10) +(−6)

   b) (−25) + (−7)

   c) (−42) + (+12)

11.2 Explore Integer Addition, 11.3 Adding Integers • MHR   109

11.4 / 11.5 Explore Integer Subtraction
Extension: Subtracting Integers

Student Textbook pp. 362–373

Key Ideas Review

Fill in the blanks with the words, diagrams, or numbers from the list. Some numbers may be used more than once.

| addition | −5 | +2 | −2 | opposite | −7 | + |

Subtraction of integers may be modelled by integer chips or by a number line.

1. To show −5 − (+2) with **integer chips**,
   - add _____ and _____ to ⊖⊖⊖⊖⊖.
   - then take away _____ ;
   - the result is _____ .

2. To show −5 − (+2) with **a number line**,
   - start at _____, then "jump" over to _____.
   - The result is _____. It is the difference and direction between the two integers.
   - Complete the number line.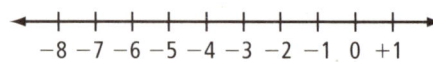

3. You can express a subtraction as the _____ of the _____ integer.
   - −5 − (+2) = −5 ___ ___
   - = ___

Example 1: Use Integer Chips to Subtract Integers

Use integer chips to find each difference.

a) (−5) − (−2)   b) (−3) − (+2)

Solution

a) ⊖⊖⊖⊖⊖ → ⊖⊖⊖⊖⊖ → ⊖⊖⊖

(−5) − (−2) = −3

b)

(–3) – (+2) = –5

Example 2: Use a Number Line to Subtract Integers

Use a number line to find the difference.

a) (–3) – (–8)   b) (–4) – (+5)

What is the result? Rewrite the expression as an addition statement.

Solution

a)
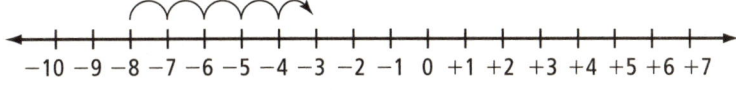

(–3) – (–8) = (–3) + (+8)
    = +5

b)
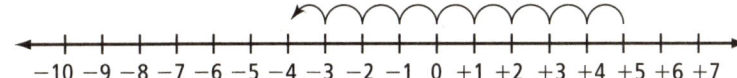

(–4) – (+5) = (–4) + (–5)
    = –9

Practise

Use integer chips to find each difference. Make a sketch. Rewrite each expression as an addition statement.

1. a) (–10) – (–7) =

b) (–6) – (–6) =

c) (–12) – (+21) =

2. Use a number line to find each difference. Rewrite each expression as an addition statement.

   a) (+4) − (+2) =

   b) (−5) − (−6) =

   c) (+10) − (+12) =

3. Each animal was asked, "What's your favourite food?"

   Frog: __ __ __ __ __ __ __ __ __ __
         +23 +6 +12 +12 +5 −28 −1 +2 −9 +36

   Cat: __ __ __ __ __ __ __ __ __ __
        −8 −9 −16 +36 +2 −51 0 0 −9 −4 −3

   To decode their answers, complete the exercise below.
   Each sum or difference represents a letter.
   Fill in the blanks with the letters that represent the numbers.

   C (−9) − (+7) =   H (−4) + (+10) =   O (+9) − (−3) =

   D (−14) − (−14) =   I (−4) + (−5) =   P (−3) − (−5) =

   E (+23) − (−13) =   L (−14) + (−14) =   S (+15) − (−8) =

   F (+12) − (+7) =   M (−5) − (+3) =   U (−69) − (−18) =

   G (−10) + (+7) =   N (+15) + (−19) =   Y (−3) − (−2) =

4. How would you show (−6) − (−6) using a number line?
   Explain. Draw a number line to help you.

Name: _____ Date: _____

11.6 Integers Using a Calculator
Student Textbook pp. 374–377

Key Ideas Review

Circle the word that will complete each sentence correctly.

1. Calculators can be used to **simplify** / **correct** integer expressions.

2. A calculator assumes that a number entered is **negative** / **positive**.

3. Your calculator may have a ⒠ key or a ⊞ key to enter **negative** / **positive** numbers.

4. To check how to enter negative integers on your calculator, use a simple **product** / **sum** that you can answer in your head.

Example: Using a Calculator to Determine a Sum

Karla's scores in four games of *Integers* are shown.

Game	Karla's Score
1	−100
2	−35
3	−64
4	137

a) Express Karla's scores as an integer expression.

b) Use a calculator to find the final score.

Use estimation to check that your answer is reasonable.

Solution

a) (−100) + (−35) + (−64) + 137

b) ⒞ 100 ⊞ ⊞ 35 ⊞ ⊟ 64 ⊞ 137 ⊜ -62

Estimate: $(-100) + (-35) + (-64) \doteq (-100) + (-100)$
$\doteq -200$
and $137 \doteq 140$

−200 + 140 = −60
The calculator answer is reasonable.

Karla's final score is −62.

> **Study Skills**
> Write the instructions for entering an integer statement on your calculator, for reference.

Practise

1. Why are flowers lazy?

To answer this riddle, follow these instructions:
- Add the four numbers that touch any letter.
- Make a chart showing the sum and the corresponding letter.
- Find your answer to the question by writing the corresponding letter for each sum.

Answer:

72	−39	45	−96	27
(E)	(H)	(R)	(C)	
−10	53	−2	−44	−32
(F)	(I)	(A)	(U)	
99	−88	−61	−19	44
(D)	(L)	(S)	(T)	
87	17	77	−23	2
(B)	(V)	(N)	(Y)	
−98	1	−57	−8	90

Sums to decode:

7, 76, −145, −126, −51, −26, 76, 4, 57, 76, 61

−55, −98, 76, −98, −11, 7, 76, 115

76, 38, 76, −97, 61, 115, −126, 61.

Apply

2. The transactions that Jocelyn and Alain made against their savings accounts during December are shown.

Jocelyn
Opening balance	1000
12/01 ATM withdrawal	100
12/06 CD purchase	12
12/15 Birthday gift deposit	200
12/16 Gift purchase for mother	70
12/22 Running shoes purchase	91

Alain
Opening balance	1200
12/03 Birthday gift deposit	350
12/08 ATM withdrawal	275
12/13 Gift purchase for mother	93
12/19 CD purchases	47
12/23 Computer part purchase	210

Who had the highest closing balance at the end of the month?

Chapter 11: Reviewing for the Test

11.1 Compare and Order Integers
page 346

1. Eight integers are shown.

   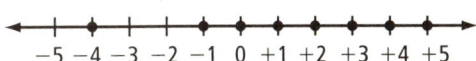

   a) Which integers are greater than 0?

   b) Which integers are less than 0?

   c) List the integers, in order, from least to greatest.

   d) Which are opposite integers?

2. Write a < or > in each ○ to compare the integers.

   a) +8 ○ −9   b) −6 ○ +6

   c) −10 ○ −7   d) 0 ○ −20

3. Write an integer sum to represent each situation.

   a) a deposit of $10, a withdrawal of $15

   b) a loss of $2, a gain of $5

   c) 8 point gain, 6 point loss

   d) sea level, 7 m below sea level, 9 m above sea level

11.2 Explore Integer Addition
11.3 Adding Integers
page 352

4. What addition statement is shown? Give the result. Describe a situation that could be represented by the statement.

   a)

   b)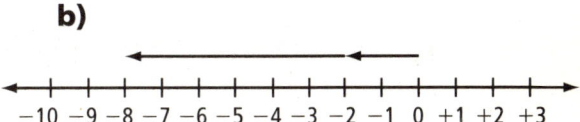

   c)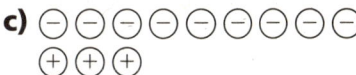

   d)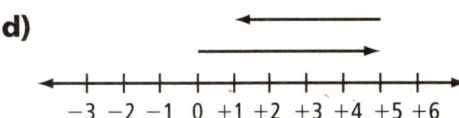

Name: _____ Date: _____

11.4 Explore Integer Subtraction
11.5 Extension: Subtracting Integers
page 362

  5. What integer difference is shown? Give the result.

a)
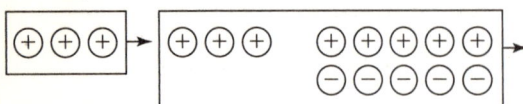

b)
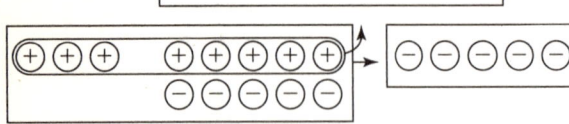

c)
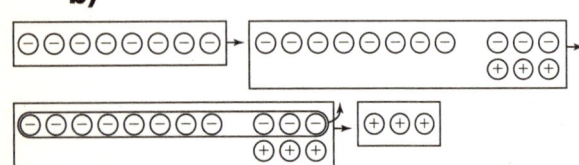

d)
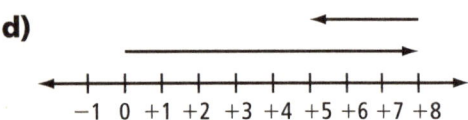

6. Complete.

a) $(-5) - (-4) = (-5) + \square$
$= \square$

b) $(-9) + (-4) = (-9) - \square$
$= \square$

c) $(+12) - (-7) = (+12) + \square$
$= \square$

11.6 Integers Using a Calculator
page 374

7. Use a calculator to evaluate each expression.

a) $12 + 10 - 9 =$

b) $(-7) - 13 + 11 =$

c) $(-4) - (-4) + 27 =$

d) $38 + (-40) - (-19) - 20 =$

f) $600 + 300 - 900 - (-700) =$

8. Evaluate each expression.

a) $127 - 245 - (-387) =$

b) $23 + (-25) - (-11) - 59 =$

c) $(-235) + (-149) - 87 + 93 =$

d) $(-39) - 39 - 39 - 39 =$

9. Find the missing numbers.

a) $-4 + 8 - \square = -12$

b) $7 + \square + 3 = -5$

c) $9 - \square - 2 = -24$

d) $\square - 25 + 37 = 0$

10. The golf scores for Carol and Juan are shown.

Carol: 3 over par, 1 under par, 1 over par, 1 under par, 2 under par

Juan: 2 over par, 1 under par, 2 over par, 1 under par, 3 under par

a) Show each person's score as an integer statement.

Carol:

Juan:

b) Who had the better score? Why?

116 MHR • Chapter 11: Integers

12.1 Variables and Expressions
12.2 Solve Equations by Inspection

Student Text pp. 386–397

Key Ideas Review

Use the words from the list to fill in the blanks.

> variable   solution   variable expression

1. _____

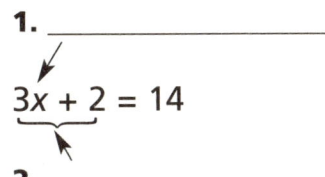

$3x + 2 = 14$ $x = 4$

2. _____   3. _____

Example 1: Model Number Phrases

Model each phrase with cups and counters or unit circles and squares.

Then, write each as a variable expression.

a) 2 times a number plus 5   **b)** 1 more than 3 times a number

Study Skills

Create a dictionary of terms for each math symbol.

+ means "plus," "more," "and"

− means "subtract," "less," "fewer"

Keep your dictionary handy when translating expressions.

Solution

a)

Let C represent the number.
$2C + 5$

b)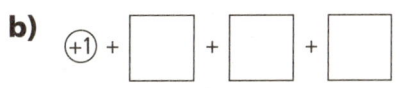

Let s represent the number.
$1 + 3s$

Example 2: Translate Models

a) Use squares and −1 circles to model the variable expression $2x - 3$.

b) Evaluate the expression for these values of x: $x = 2$, $x = 3$, and $x = 4$.

variable expression
- contains variables and operations with numbers
- $C + 3$ and $2C$ are variable expressions

Solution

a) ☐ + ☐ + (−1) + (−1) + (−1)

b) Method 1: Use Squares and –1 Circles

For $x = 2$:
$2 + 2 + (-1) + (-1) + (-1)$
$= 4 + (-3)$
$= 4 - 3$
$= 1$

For $x = 3$:
$3 + 3 + (-1) + (-1) + (-1)$
$= 6 + (-3)$
$= 6 - 3$
$= 3$

For $x = 4$:
$4 + 4 + (-1) + (-1) + (-1)$
$= 8 + (-3)$
$= 8 - 3$
$= 5$

Method 2: Substitute Into the Expression $2x - 3$

For $x = 2$:
$2 \times 2 + (-3)$
$= 4 - 3$
$= 1$

For $x = 3$:
$2 \times 3 + (-3)$
$= 6 - 3$
$= 3$

For $x = 4$:
$2 \times 4 + (-3)$
$= 8 - 3$
$= 5$

Example 3: Solve Equations

Find a number that is the solution for $2x + 5 = 13$.

> **equation**
> - shows that two expressions are equal
> - the expressions are separated by an equals symbol
> - $C + 3 = 2C$ and $4m = 8$ are equations

Solution

Method 1: Use Inspection

$2x + 5 = 13$
$8 + 5 = 13$
So $2x = 8$.
$2 \times 4 = 8$

The solution is $x = 4$.

> **solution**
> - a number that makes an equation true
> - $m = 2$ is the solution to $4m = 8$

Method 2: Test for Solutions

Which of the numbers 2, 3, or 4 is the solution to $2x + 5 = 13$?

Try 2:
$2(2) + 5$
$= 4 + 5$
$= 9$
$x = 2$ does not work.

Try 3:
$2(3) + 5$
$= 6 + 5$
$= 11$
$x = 3$ does not work.

Try 4:
$2(4) + 5$
$= 8 + 5$
$= 13$
$x = 4$ works.

The solution is $x = 4$.

Name: _____ Date: _____

Practise

1. Write variable [expressions] to model each diagram.

   a)

   b)

2. Each cup in question 1 contains seven counters. How many counters are there for each model?

   a)

   b)

3. Model each variable expression.

   a) $4k + 2$

   b) $2M - 6$

4. For each expression in question 3, substitute the value 5 for the variable and evaluate.

   a)

   b)

5. Write variable equations for each model. How many counters are there in each cup?

   a)

   b)

6. Which of the numbers 2, 3, or 4 is a solution to the following equations?

   a) $3x = 12$

   b) $2b - 8 = -2$

   c) $12m + 8 = 32$

Apply

7. Show each phrase with a variable expression, using addition, subtraction, or multiplication.

   a) $5 *less* than an unknown price

   b) the *product* of 4 and the number of tickets sold

   c) the perimeter *increased* by 7 m

   d) *half* the regular price

8. Let each expression in question 7 equal 20. Find a number that is a solution to the equation.

   a)

   b)

   c)

   d)

12.1 Variables and Expressions, 12.2 Solve Equations by Inspection • MHR   119

12.3 Model Patterns With Equations

Student Text pp. 398–403

Key Ideas Review

- The items in a pattern can be translated into equations, using numbers or variables.
- An equation for a pattern can be explained in terms of the pattern.

Diagram 1   Diagram 2   Diagram 3

d represents the diagram number.
Fill in the blanks to complete the pattern.

$2d + \_\_ = 5 \quad 2\_\_ + \_\_ = 7 \quad 2\_\_ + \_\_ = \_\_$

Complete each equation.
- You can write an equation in different ways.
  $2n - 1 = 3, \_\_ + 2n = 3, 3 = 2n - \_\_, 3 = _____$ all mean the same thing.

Practise

1. a) Complete the table for the pattern.

Picture Number	Number of Dots	Equation
1	2 × 1 centre dot + __ end dots	2 × __ + 2 = 4
2	2 × __ centre dot + __ end dots	2 × __ + 2 = __
3	__ × __ centre dots + __ end dots	__ × __ + __ = __
⋮		
7	__ × __ centre dots + __ end dots	__ × __ + __ = __

b) Let p represent the picture number. Write an equation to model the picture with 17 centre dots. Explain what your equation means.

2. a) Use equations to model the first three diagrams in this pattern.

Diagram 1:

Diagram 2:

Diagram 3:

b) Write an equation for the diagram with 31 toothpicks in two ways.

c) How many triangles are there with 31 toothpicks?

3. a) Draw the first three diagrams with marbles in a pattern described by $4 + 3n$.

b) Use an equation to model the diagram with 19 marbles. Explain the meaning of your equation. Which diagram has 19 marbles?

12.4 Solve Equations by Systematic Trial

Student Text pp. 404–409

Key Ideas Review

Fill in the blanks with the words from the list.

> variable   greater   substitute   lesser

1. To solve equations by systematic trial, _____ values for the _____.

2. If your answer is less than the desired number, then use a _____ value.

   If your answer is greater than the desired number, then use a _____ value.

   Keep testing values until you get the correct answer.

Example: Model and Solve Pattern Problems

a) Write an equation for the diagram with 28 dots.

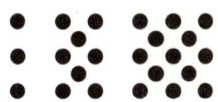

b) Solve your equation. State what your solution means.

Solution

a)

Diagram Number	Number of Dots	Pattern
1	3	$5 \times 1 - 2$
2	8	$5 \times 2 - 2$
3	13	$5 \times 3 - 2$
d	28	$5 \times d - 2$

The equation is $5d - 2 = 28$.

Hint

The number of dots increases by five in the next diagram.

Show this as multiplying the diagram number by 5.

b) Use systematic trial.

Try $d = 5$:
$5 \times 5 - 2$
$= 25 - 2$
$= 23$ **Too small.**

Try $d = 7$:
$5 \times 7 - 2$
$= 35 - 2$
$= 33$ **Too large.**

Try $d = 6$:
$5 \times 6 - 2$
$= 30 - 2$
$= 28$ **Correct.**

The solution is $d = 6$.

In Diagram 6, there will be 28 dots.

Name: _____ Date: _____

Practise

1. Study the pattern of coins.

Diagram 1   Diagram 2   Diagram 3

a) Copy and complete the table.

Number of Quarters	Coins	Value of Coins (¢)
1	1 dime + 1 quarter	$10 \times 1 + 25 \times 1$
2		
3		
q		

Hint

1 dime = 10¢   1 quarter = 25¢ $1.60 = 160¢

b) Write an equation for the diagram in which the value of the coins is equal to $1.60.

c) Solve your equation. State what your solution means.

Apply

2. Answer each question to develop an expression to model the number pattern 91, 85, 79, 73, ...

a) How does each number compare to the one before it?

b) Use the variable x to write an expression for a step in the pattern.

3. Which step (x) gives you the number 1 in the pattern in question 2? Substitute $x = 15$ into your expression from question 2. Is the result close to 1? Try different step numbers until you get 1 as an answer.

4. Solve each equation by systematic trial.

a) $80 = 8 + 6p$

b) $2q - 7 = 37$

5. A weighing tray has a mass of 2 kg. Bryce places 3-kg bags of oranges on the tray. How many bags are on the tray when the total mass is 41 kg? Use systematic trial.

12.5 Model With Equations

Student Text pp. 410–415

Example: Model With an Equation and Solve

Janie collected $13.50 in coins for the walk-a-thon. She has $12 in one-dollar coins and the rest is in quarters. How many quarters does she have?

Solution

Method 1: Use an Equation and Systematic Trial

Let q represent the number of quarters.
The quarters have a value of $25q$ in cents.
An equation that models the situation is $1200 + 25q = 1350$.

Try $q = 4$:
$1200 + 25 \times 4$
$= 1200 + 100$
$= 1300$ **Too low.**

Try $q = 6$:
$1200 + 25 \times 6$
$= 1200 + 150$
$= 1350$ **Correct.**

The solution is $q = 6$. Janie has six quarters.

> **Hint**
> Express the equation in cents.
> $12 = 1200¢$
> $13.50 = 1350¢$

Method 2: Use an Equation and Work Backward

An equation that models the situation is $1200 + 25q = 1350$.
There are 1200¢, so the left-over money is $1350 - 1200 = 150$
The left-over money is 150¢, so $25q = 150$
$q = 6$

Janie has six quarters.

> **Hint**
> There are four quarters in one dollar.
> There are two quarters in 50¢.

Practise

1. Write each sentence as an equation, then solve.

   a) Nine more than triple a number is 33.

   b) The price of ten identical baseball caps plus $4 is $34.

2. What situations could these equations describe?

   a) $32 + 3h = 47$ b) $17 = 41 - 4y$

Apply

3. Three songs on a CD are each six minutes long. The rest are each four minutes long. How many songs are on the CD if it is one hour and six minutes long? Use an algebraic equation.

Chapter 12: Reviewing for the Test

12.1 Variables and Expressions
12.2 Solve Equations by Inspection
page 386

1. a) Each box contains the same number of balls. Which expression models this diagram?

   □ + □ + □ + □ + □ − ∘∘

   A $4 - 5p$ B $5 - 4B$

   C $5x - 4$ D $4b - 5$

   b) Each box contains six balls. How many balls are there in total?

2. How many balls are in each box?

12.3 Model Patterns With Equations
page 398

3. a) Model the number of toothpicks in each item of the pattern using an algebraic expression.

   b) Draw the fifth item in the pattern.

   c) How many squares are there when there are 31 toothpicks?

12.4 Solve Equations by Systematic Trial
page 404

4. Solve each equation by systematic trial.

   a) $83 - 4p = 35$

   b) $128 = 3z + 5$

12.5 Model With Equations
page 410

5. Cosette babysits for $8/hour. She was paid $72 for two weeks. How many hours did she work?

6. Kaye paid $12.50 for four markers and one notebook. The notebook cost $9.50. How much was each marker?

Name: _____   Date: _____

13.1 Explore Transformations

Student Text pp. 428–433

Key Ideas Review

Use the words from the list to label the diagrams and fill in the blanks.

reflection	translation	rotation	image
mirror line	translation arrow	angle of rotation	
turn centre	transformation	rotation	

A: _____   D: _____   F: _____

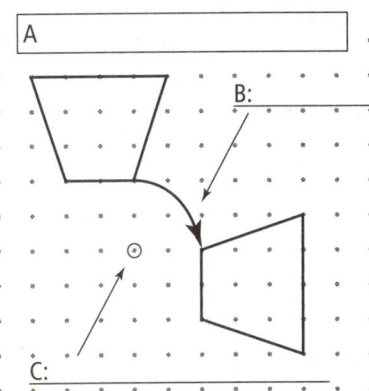

B: _____

C: _____

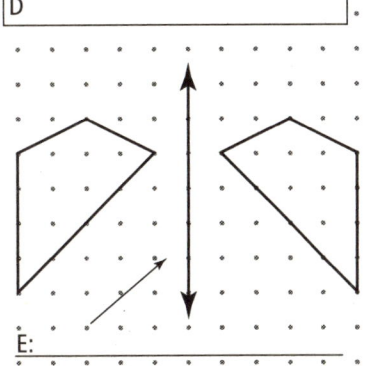

E: _____

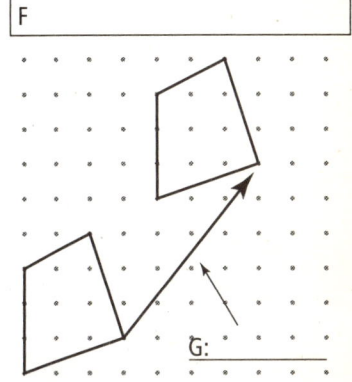

G: _____

- Each picture shows a different example of a

  _____.

- The first shape is the original and the second shape is

  the _____.

Literacy Connections

There are other words for the different transformations.

A translation is also called a slide.

A reflection is also called a flip.

A rotation is also called a turn.

Example: Identify Transformations

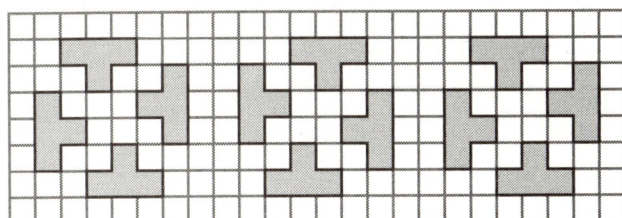

A tetromino is four congruent squares joined along whole edges.
What transformations are in this frieze pattern?

frieze pattern
- a design pattern that repeats in one direction

Solution

Group 1: Each tetromino is a rotation of the others.
Group 2: Group 2 is a reflection of Group 1.
Group 3: Group 3 is a translation of Group 1, 16 units to the right.

13.1 Explore Transformations • MHR 125

Practise

1. Name the type of transformation that relates each pair of figures. Show the translation arrow, mirror line, turn centre, or angle of rotation.

   a)

   _____

   b)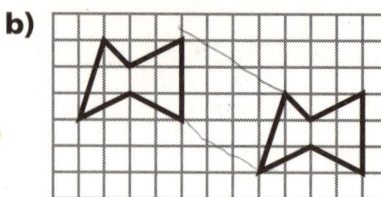

   _____

   c)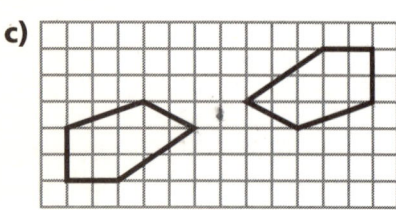

   _____

   d)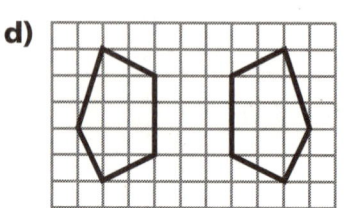

   _____

2. Perform each transformation.

   a) Translate the shape.

   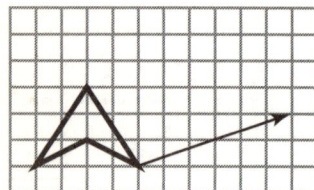

   b) Reflect the shape.

   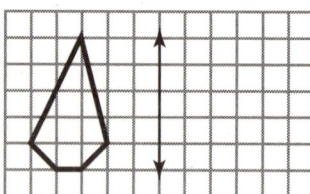

   c) Rotate the shape a quarter-turn.

   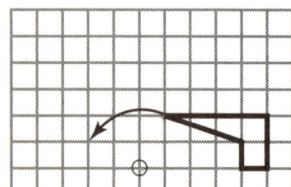

   d) Translate the shape.

   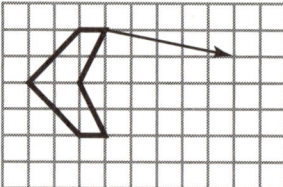

3. Create a frieze pattern for a border of a room. Use all of the transformations at least once.

Name: _____   Date: _____

13.3 Extension: Translations on a Coordinate Grid

Student Text pp. 436–441

Example: Translate a Figure

Quadrilateral ABCD is translated 2 units left (along the x-axis) and 3 units up (along the y-axis).

Draw its translation image, quadrilateral A'B'C'D'.

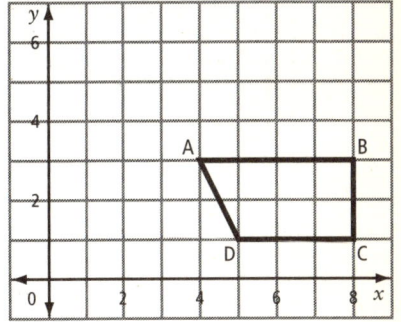

Solution: Use a Diagram

Draw the diagram on grid paper.

Count squares to move each vertex 2 units left and 3 units up. Label the vertices A', B', C', and D'.

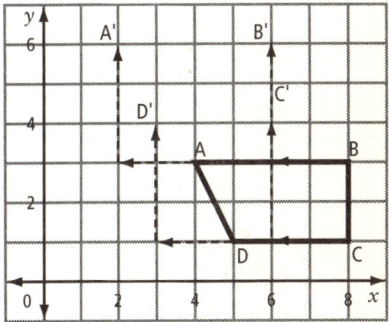

Join A', B', C', and D' to form the translation image.

Check by creating quadrilateral ABCD on grid paper.

Then translate ABCD on a coordinate grid.

Trace the final position of A'B'C'D'.

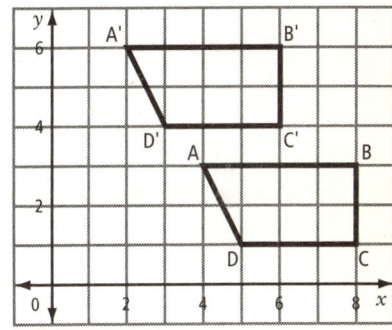

Practise

1. Perform each translation.

a) Translate the shape right 3 then up 2.

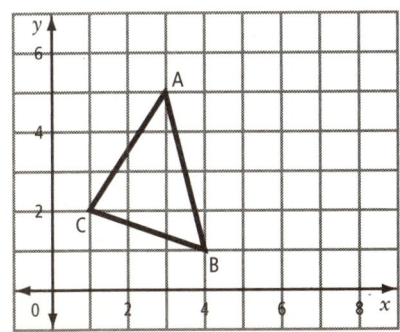

b) Translate the shape to match the arrow.

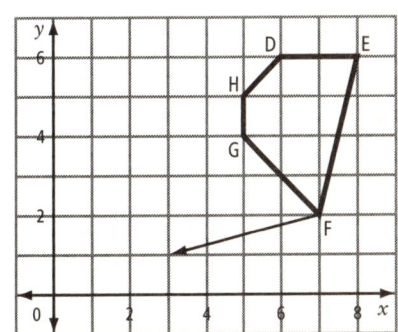

13.3 Extension: Translations on a Coordinate Grid • MHR 127

2. Describe the translation that moves each figure onto its image.

a)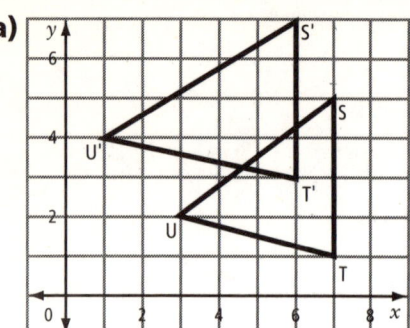

b)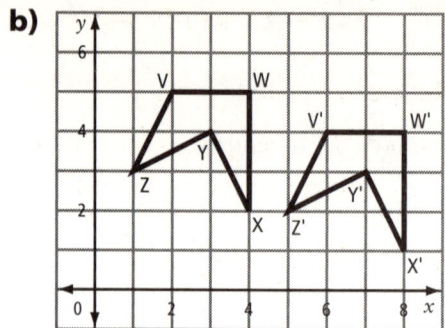

3. Find the five triangles that are translations of the shaded triangle. Unscramble their letters to spell the missing word.

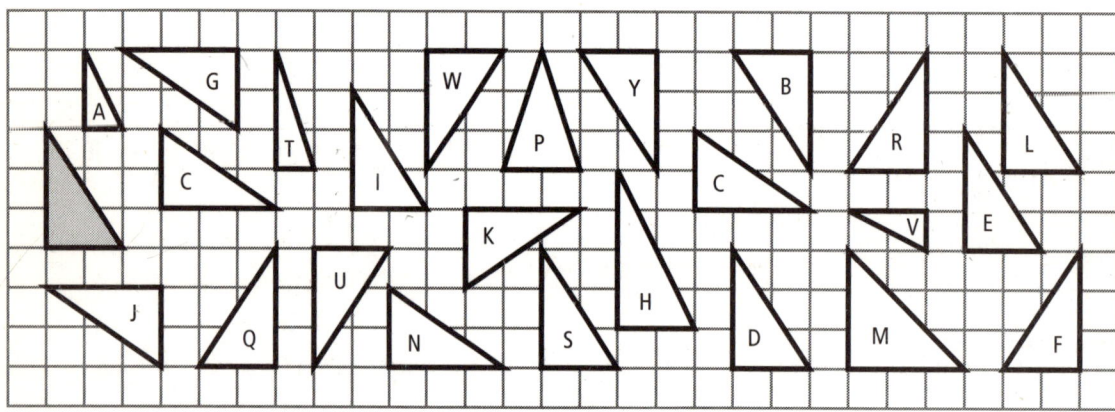

Another word for translation is ___ ___ ___ ___ ___.

4. △XYZ is translated 7 units right and 4 units up. Then it is translated 12 units left and 2 units down. What single translation would move △XYZ onto △X'Y'Z' directly? Use diagrams to illustrate your answer.

128   MHR • Chapter 13: Geometry of Transformations

13.4 Identify Tiling Patterns and Tessellations

Student Text pp. 442–445

Key Ideas Review

1. A tiling pattern or tessellation is a pattern that covers a plane without overlapping or leaving gaps.   True   False

2. All irregular figures tile the plane.   True   False

3. Only three regular figures tile the plane. Identify these figures.

A. _____   B. _____   C. _____

Example: Explore Tiling with Regular Octagons

Can you create a tessellation based on regular octagons and squares?

Solution

Regular octagons do not tile the plane.
Regular hexagons, squares, and equilateral triangles do tile the plane.
Start placing regular octagons on the plane.

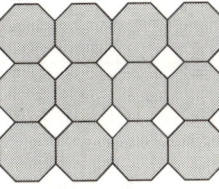

Among every four octagon tiles, there is a square.
Create a tile that combines the octagon and the square.

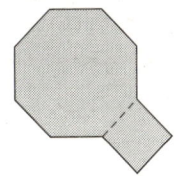

Start placing the new figure on the plane.

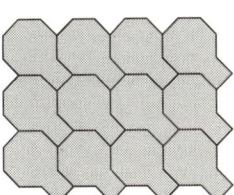

This tiling covers the plane.
It is a tessellation.

Practise

1. Use the given shape to tile the plane.

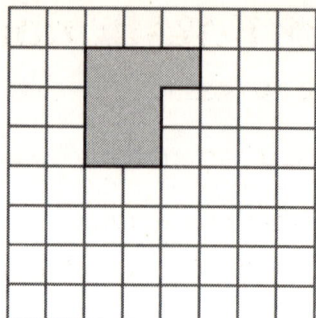

Hint

Trace the shape on paper and then cut it out.

Use your shape to fill the grid.

Trace your cutout each time you place it on the grid.

2. This shape is popular for paving stones. Show another way to tile the plane using the shape.

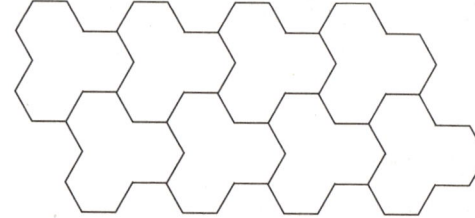

Making Connections

Look around your home; where do you see tiling patterns?

Make a drawing of one of the patterns.

What transformation of a basic shape occurs?

3. Which shapes will tile the plane? Explain your answers.

a)

b)

c)

d)

13.5 Construct Translational Tessellations
13.6 Construct Rotational Tessellations

Student Text pp. 448–455

Key Ideas Review

Circle the correct word to complete each sentence.

1. In a **rotational** / **translational** tessellation, every tile is a translation of the other tiles.

2. In a **rotational** / **translational** tessellation, the tiles can be translations or rotations of each other.

Practise

1. Identify which tessellations are translational and which are rotational:

   a) 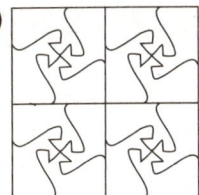   b) 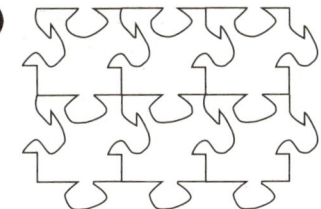   c)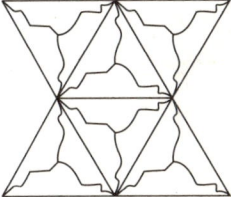

   d)    e) 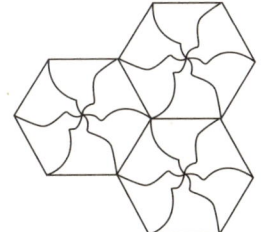   f)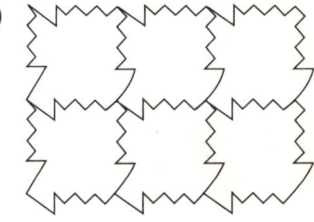

2. Design a translational tessellation based on a parallelogram.

3. Design a rotational tessellation based on an equilateral triangle.

Chapter 13: Reviewing for the Test

13.1 Explore Transformations
page 428

1. Which two transformations are shown?

   A Reflection

   B Rotation

   C Translation

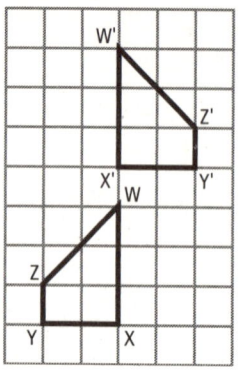

13.3 Extension: Translations on a Coordinate Grid
page 436

2. Describe the translation.

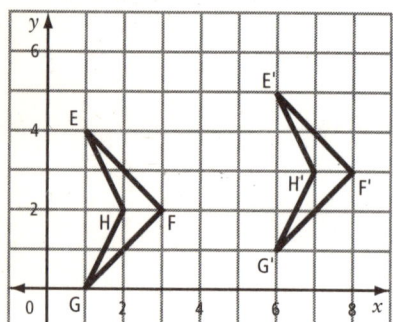

3. Rotate quadrilateral EFGH, in question 2, 180° around the point (3, 3). Where is E'? Show your work on the grid in question 2.

13.4 Identify Tiling Patterns and Tessellations
page 442

4. Will this shape tile the plane? Explain your answer.

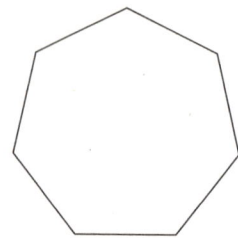

5. Use the given shape to tile the plane.

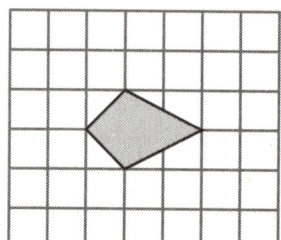

13.5 Construct Translational Tessellations
13.6 Construct Rotational Tessellations
page 448

6. Is the tessellation in question 5 rotational or translational? Explain.

7. Identify each tessellation as rotational or translational.

   a)    b)

Name: _____ Date: _____

Summer Tune-up: A Note to Parent/Guardian

This section of the workbook contains several practical and useful activities in the context of summer fun and outings. The focus is on review and practice of grade 7 concepts and skills in preparation for grade 8. The activities have been constructed so that they can be done anywhere: in the back seat of a car, on the beach, or under a shady tree. Several different kinds of puzzles are included. No special tools other than a pencil, paper, ruler, and calculator are required. Learning doesn't have to end with the school year!

If you have access to a computer, additional features are available at www.mcgrawhill.ca. Log on and navigate to the grade 7 workbook page then follow the links.

A Day at the Air Show

Charlie and Patty are attending an air show with their parents.

The Snowbirds

The stars of the air show are the famous Canadian Snowbirds.

One of their flying formations is shown.

1. List three pairs of congruent triangles.

2. Measure the sides of quadrilateral ACGE. Identify the quadrilateral. Explain how you know.

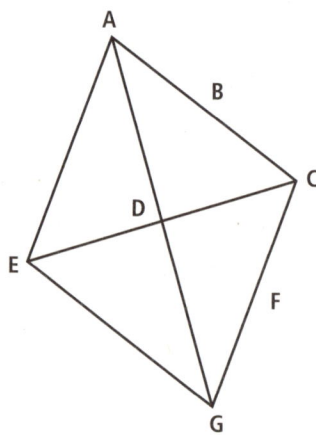

Kites, Nets, and Patterns

Charlie purchased a kite in the shape of an aircraft at one of the air show stands. The kite is the net of a three-dimensional figure.

3. Calculate the area of the kite to the nearest square centimetre. Show how you got your answer.

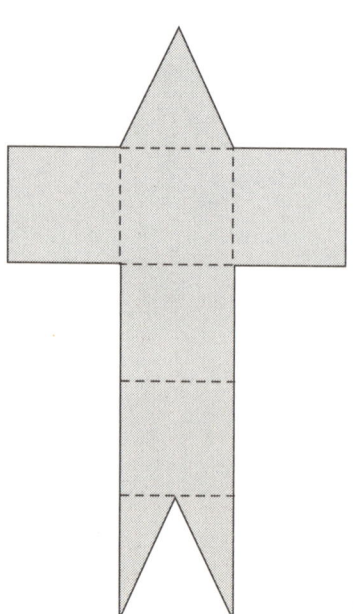

134 MHR • Preparing for Grade 8

4. Trace the shape onto a sheet of paper. Fold it into a three-dimensional figure. What figure is it?

5. Calculate the volume of the three-dimensional figure from question 4.

6. Copy Charlie's kite onto graph paper but make the sides twice as long. Fold the net into a three-dimensional shape. Predict how many of the figure from question 4 will fit into the larger one. Check by calculating the volume of the larger figure.

7. Repeat the steps in question 6 but make the sides three times as long. Predict how many of the figure from question 4 will fit into the larger one. Check by calculating the volume of the larger figure.

8. Predict how many of the figure from question 4 will fit into a figure made from a net with sides four times as long. Use the pattern in the numbers from questions 5, 6, and 7 to help you.

Aerobatics

One popular air show manoeuvre is the "Lazy Eight". Draw the figure on graph paper.

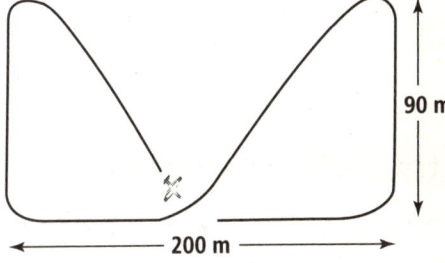

9. Estimate the perimeter of the figure. Explain how you got your answer.

10. Estimate the area of the figure. Explain how you got your answer.

Name: _____   Date: _____

A Baseball Game

Tim, Anwar, and Tadeuz went to a baseball game to watch their favourite team. They have been following the team's statistics.

Player	AG	G	AB	R	H	HR	RBI
10	31	88	273	29	67	3	22
11	30	143	505	013	140	33	108
13	31	109	374	42	101	4	39
14	24	151	566	99	158	24	84
19	22	85	282	35	64	8	34
22	28	141	577	103	175	10	45
25	23	159	608	87	167	23	100
29	31	75	299	51	67	15	45
31	24	74	265	41	82	15	58
33	28	124	466	64	114	18	70
37	26	90	312	48	86	13	45
42	31	96	265	33	68	8	37
44	24	54	192	20	53	4	23
49	25	55	150	16	27	2	15
53	35	45	127	8	28	3	22
55	24	42	112	14	20	3	13
58	29	23	78	9	18	1	2
61	23	15	46	4	12	0	6
65	31	24	38	2	5	0	2
68	24	7	14	1	2	0	0
74	31	13	12	3	1	0	1
76	25	2	6	0	0	0	0
77	30	2	6	0	1	0	0
82	34	2	4	0	0	0	0
88	24	2	2	0	0	0	0
96	27	1	1	1	1	0	0

AG = age of the player
G = number of games played
AB = number of times at bat
R = number of runs made
H = number of hits
HR = number of home runs
RBI = number of runs batted in

Use your calculator to learn some quick facts about this team:

1. Which player has the best ratio of home runs to games played?

2. Which player has the best ratio of home runs to times at bat?

3. What is the mean number of games played per player?

136   MHR • Preparing for Grade 8

4. What is the median number of times at bat for a player?

5. What is the mode for the age of the players?

Age Bias?

Tim claims that the makeup of the team is biased in favour of younger players.

6. Complete the table for the age data.

Age	Tally	Frequency
22		
23		
24		
25		
26		
27		
28		
29		
30		
31		
32		
33		
34		
35		

7. Complete the stem-and-leaf plot for the data.

Stem (tens) | Leaf (ones)

8. Was Tim correct about the age bias? Explain your answer.

Making Predictions

9. Player #10 is at bat. What is the probability that he will get a hit?

10. Player #76 is next to bat. What is the probability that he will get a hit?

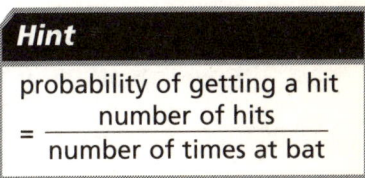

Hint

probability of getting a hit = $\dfrac{\text{number of hits}}{\text{number of times at bat}}$

11. If you were the team's coach, which player, #10 or #76, would you use more often at bat? Explain.

Searching For Geometers

Clues for geometric terms are shown below.
Circle each term in the word search pattern.

```
N  O  I  T  A  L  E  S  S  E  T  K  D  Q  T
S  Q  U  A  R  E  F  L  E  C  T  I  O  N  S
O  A  E  A  X  A  T  H  R  X  O  X  E  Z  T
H  F  N  G  T  L  N  H  E  Z  C  U  B  E  S
G  R  A  L  I  M  I  S  E  X  R  N  N  N  F
W  S  L  B  J  S  E  P  L  G  A  Z  P  E  Q
O  J  P  Z  X  I  A  P  N  A  M  G  C  H  K
U  V  M  A  W  R  R  O  T  A  T  I  O  N  S
H  M  G  H  T  P  C  V  H  C  G  I  E  N  L
R  Z  Z  R  K  I  N  Z  O  A  W  Y  O  J  D
Z  F  S  F  Z  R  A  C  J  T  D  V  X  N  R
N  C  O  N  E  V  G  L  D  I  M  A  R  Y  P
```

1. Name for figures that are the same shape and size.
2. A six-sided polygon.
3. Another name for a flip.
4. Figures that are the same shape but different sizes.
5. A quadrilateral with all sides equal and and all angles 90 degrees.
6. A quadrilateral with one pair of parallel sides.
7. A three-dimensional figure with rectangular faces, whose base and top are parallel polygons.
8. A two-dimensional pattern for a three-dimensional figure.
9. A three-dimensional figure with all square faces.
10. Another name for a turn.
11. A pattern that covers a plane exactly.
12. Another name for a slide.
13. A three-dimensional figure with polygon base and triangular sides meeting at one point.
14. The three-dimensional figure that a safety pylon resembles most closely.

Name: _____ Date: _____

Road Trip to Ottawa

Victor and Leela Jung live in Kingston with their mother, Mary.

They are planning a day trip to Ottawa.

Mary will be driving, Victor will plan the route, and Leela will organize the food and beverages.

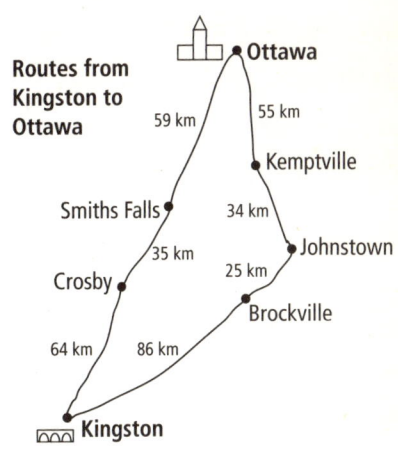

Routes from Kingston to Ottawa

Preparing for the Trip

1. Use the road map to outline two different routes from Kingston to Ottawa.

2. It takes 50 g of drink crystals to make 1 L of fruit drink. How many grams are needed to make 3.5 L?

3. The Jungs have allowed a budget of $40.00 for sandwiches, drinks, and fruit. Leela bought 10 cans of juice at 75¢ each and a dozen apples for $2.50. Leela estimates it will cost $1.50 to make a sandwich. Complete the table. Add more rows to the table if necessary.

Number of Sandwiches	Cost of Juice and Apples ($)	Total Cost ($)
1	10 × 0.75 + 2.50 = 10.00	1 × 1.50 + 10.00 = 11.50
2		
3		
s		

4. Write the equation for the total cost of the food.

5. How much would it cost to make 25 sandwiches?

6. How many sandwiches could Leela make if her budget were $55.00?

Preparing for Grade 8 • MHR 139

The Road Map

7. Calculate the distance for each route from Kingston to Ottawa.

8. The shortest route from Kingston to Ottawa is along a four-lane highway. The speed limit is 100 km/h. The longest route is along a two-lane highway. The speed limit is 70 km/h. If Mary drove at the speed limit, estimate how long it would take her to get to Ottawa along each route.

9. The routes from Kingston to Ottawa form a triangle. What kind of triangle is formed? Explain your answer.

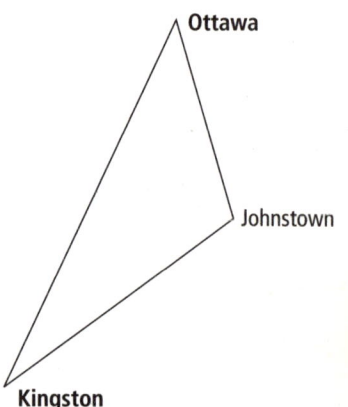

Fuel Consumption

The Jungs' car uses 5.8 L of fuel for every 100 km when driving at 100 km/h and 7.2 L when driving at 70 km/h.

10. How much fuel is required for each route to the nearest tenth of a litre?

11. If fuel costs 82.9¢ per litre, what is the fuel cost for each route to the nearest dollar?

Calling All Triangles: Telephone Scramble

Select one of the letters in each box to reveal four kinds of triangles.

DEF	QZ	TUV	GHI	JKL	ABC	TUV	DEF	PRS	ABC	JKL
3	0	8	4	5	2	8	3	7	2	5

GHI	PRS	MNO	PRS	ABC	DEF	JKL	DEF	PRS
4	7	6	7	2	3	5	3	7

PRS	GHI	GHI	GHI	TUV	ABC	MNO	GHI	JKL	DEF	DEF
7	4	4	4	8	2	6	4	5	3	3

PRS	ABC	ABC	JKL	DEF	MNO	DEF
7	2	2	5	3	6	3

The triangles fit the following definitions but not necessarily in the order given.

1. Two sides equal

2. All sides equal

3. Contains one right angle

4. No sides equal

Name: _____   Date: _____

Data Scramble

Unscramble the letters to reveal five terms used in data management.

A tool for counting data:
CHLRATTYLA   ____ ____ ____ ____ ____   ____ ____ ____ ____ ____

Way to display data using sectors of a circle:
ELRRGHAICCP   ____ ____ ____ ____ ____   ____ ____ ____ ____ ____

Most common data:
EMDO   ____ ____ ____ ____

Half above, half below:
ENADIM   ____ ____ ____ ____ ____ ____

Way to display data using bars:
PAHRGRAB   ____ ____ ____   ____ ____ ____ ____

Measure of central tendency, average:
MNAE   ____ ____ ____ ____

Way to display data using small pictures:
PHATGPORCI   ____ ____ ____ ____ ____ ____ ____ ____ ____

Number of a particular item in a set of data:
YQUFCNERE   ____ ____ ____ ____ ____ ____ ____ ____ ____

Name: _____   Date: _____

Licence Plate Game

Play the Licence Plate Game on a long road trip with another passenger.

Rules

Watch for 20 oncoming cars. This is one round of the game.

Look for licence plates whose last three or four figures are numbers.

Record the last two numbers.

A win is either: at least two of your cars having matching numbers; or there being no match.

Use a table to record your data.

Car Number	Last Two Numbers	Car Number	Last Two Numbers
1		1	
2		2	
3		3	
4		4	
5		5	
6		6	
7		7	
8		8	
9		9	
10		10	
11		11	
12		12	
13		13	
14		14	
15		15	
16		16	
17		17	
18		18	
19		19	
20		20	

Number Facts

1. How many possible combinations are there for the last two numbers of a licence plate?

2. Calculate the probability of a match.

3. Calculate the probability of no match.

4. Which has the higher probability: a match or no match?

5. Play the game for ten more rounds. Is your answer to question 2 still the same after each round? Give reasons for the differences.

Name: _____   Date: _____

Number Generator

Match the items in Column A with their correct answer in Column B.

A

1. 2 − 5
2. 40% as a decimal
3. $\frac{1}{2} - \frac{1}{3}$
4. 10% of 50
5. 4^2
6. Half of 58
7. Lowest common multiple of 6 and 9
8. Useful rule for the order of operations
9. $\sqrt{64}$
10. Lowest common denominator of $\frac{1}{2}$ and $\frac{1}{5}$
11. 25% as a fraction in lowest terms
12. 3 × 3 × 3 × 3 in exponential form

B

a) BODMAS
b) 29
c) $\frac{1}{6}$
d) $\frac{1}{4}$
e) 3^4
f) 0.40
g) 16
h) 10
i) −3
j) 5
k) 8
l) 18

Magic Squares

The numbers in every row, column, and diagonal have the same sum. Fill in the missing numbers.

		4
	5	3
6		

		22
	4	
−14		10

144   MHR • Preparing for Grade 8

A Canoe Trip

Erika, Mandy, Sabina, and Soong are planning a three-day canoe trip in Algonquin Park.

Drop-off will be at 8 a.m. at Canoe Lake on Day 1, and pickup will be at 9 p.m. at Canoe Lake on Day 3.

Getting Ready

Erika is in charge of researching the weather forecasts over the next three days.

Day 1: no precipitation, morning low of −4°C, daytime high of 13°C

Day 2: morning precipitation, morning low of 0°C, daytime high of 14°C

Day 3: no precipitation, morning low of +2°C, daytime high of +16°C

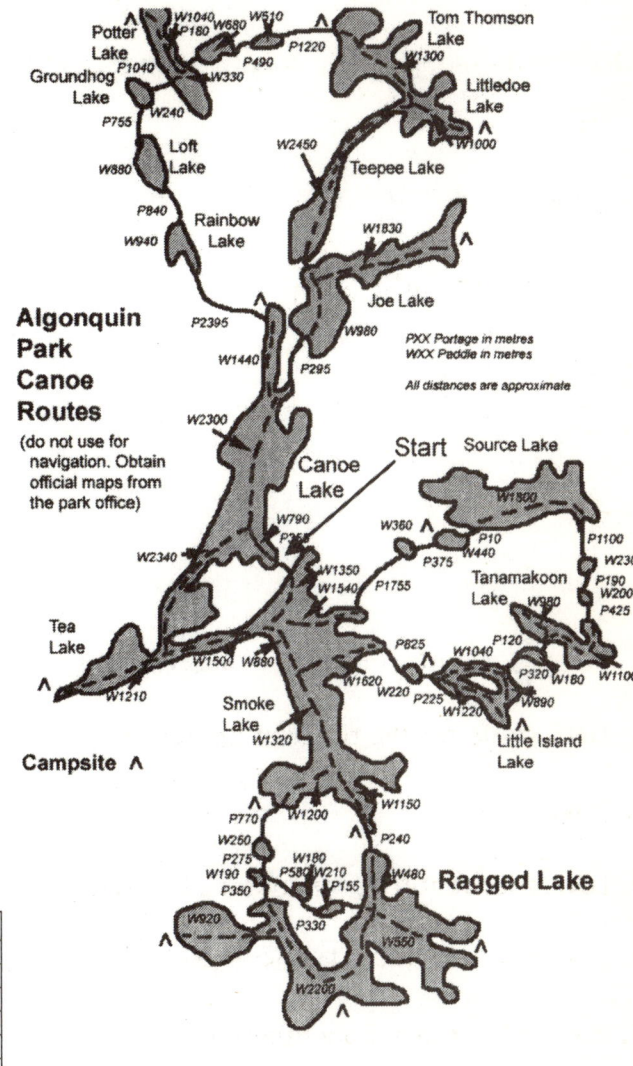

1. Show the changes in temperature for each day on the thermometers.

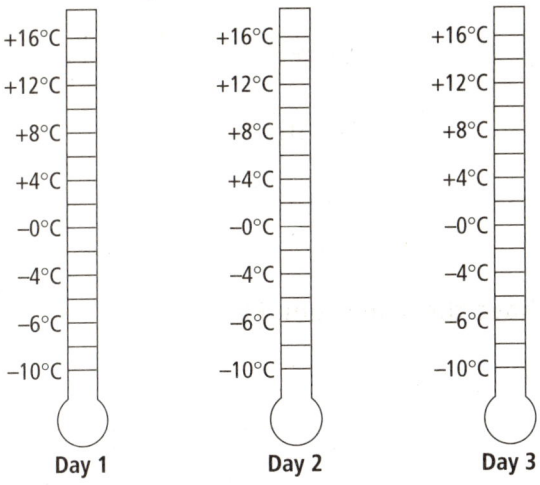

2. Write addition sentences to represent the change in temperature for each day.

   Day 1:

   Day 2:

   Day 3:

Name: _____   Date: _____

3. Write each addition sentence in question 2 as a subtraction sentence.

   Day 1:

   Day 2:

   Day 3:

4. Which day had the greatest change in temperature?

Purchasing the Food

Mandy is in charge of organizing the food.

She uses a grocery store flyer to assemble a sample menu.

Discount Foods
Bulk Foods
Powdered milk	$0.28/100 g
Juice crystals	$0.32/100 g
Cereals	$0.72/100 g
Dried fruits	$0.98/100 g
Crackers/cookies	$0.66/100 g
Spreads	$1.12/100 g
Dried mixes	$1.04/100 g
Snack foods	$0.48/100 g

Produce
Oranges	$0.25 ea
Bananas	$0.12 ea
Apples (all kinds)	$0.30 ea
Broccoli	$0.99 head
Lettuce	$0.89 head

Menu (per person)
Breakfast	Powdered Milk	80 g
	Orange Crystals	30 g
	Granola	200 g
	Raisins	50 g
	Banana	1
Lunch	Cranberry Crystals	50 g
	Crackers	110 g
	Cheese Spread	50 g
	Jam	100 g
	Cookies	90 g
	Orange	1
Dinner	Grape Crystals	75 g
	Dried Stew Mix	150 g
	Biscuit Mix	80 g
	Corn Chips	120 g
	Chocolate Wafer	70 g
	Apple	1

5. a) How many breakfasts will be eaten over the three days? Explain.

   b) How many lunches will be eaten over the three days? Explain.

   c) How many dinners will be eaten over the three days? Explain.

Name: _____   Date: _____

6. Complete the tables that help Mandy determine her shopping list for the trip.

Breakfast	Quantity for One Person's Meal	Quantity for Group	Quantity for Trip	Cost
Powdered milk	80 g	4 × 80 g = 320 g	2 × 320 g = 640 g	
Juice crystals-orange				
Cereal-granola				
Dried fruit-raisins				
Bananas				
Total Cost				

Lunch	Quantity for One Person's Meal	Quantity for Group	Quantity for Trip	Cost
Juice crystals-cranberry	50 g	4 × 50 g = 200 g	3 × 200 g = 600 g	6 × $0.32 = $1.92
Snack foods-crackers and cookies				
Spreads-cheese and jam				
Oranges				
Total Cost				

Dinner	Quantity for One Person's Meal	Quantity for Group	Quantity for Trip	Cost
Juice crystals-grape	50 g	4 × 50 g = 200 g	3 × 200 g = 600 g	6 × $0.28 = $1.68
Dried mixes-stew and biscuit				
Snack foods-corn chips and chocolate wafers				
Apples				
Total Cost				

7. How much is each person's share of the cost of the food to the nearest dollar?

Navigation and Time Log

Sabina and Soong are in charge of planning the canoe route. The trip will begin and end at the south end of Canoe Lake. Sabina plans to paddle at an average speed of 3 km/h and to portage at an average speed of 1 km/h.

8. Plan a route that will take three days to complete. Six hours per day are spent canoeing and portaging. You must camp at one of the designated sites. Plan meal times and locations for each day. Prepare a log showing the distances and estimated times required each day. What is the total distance covered on the trip?

Mobile Polyhedra

The following are nets of polyhedra.

A.

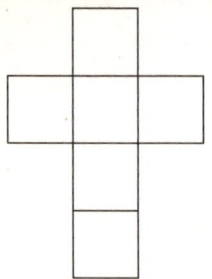

All sides measure 6 cm.

B.

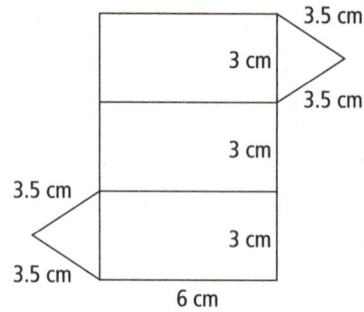

C.

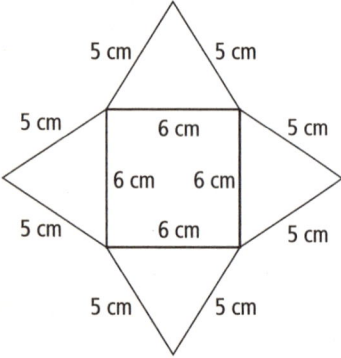

D.

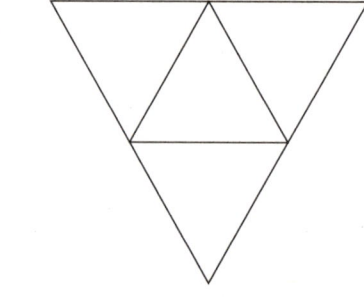

All sides measure 6 cm.

1. Copy the nets onto graph paper using the measures given. Cut out then fold the nets to form the polyhedra. Secure the edges with tape. Create a mobile from your polyhedra. Attach a piece of string to each figure with tape. Tie each string onto a stick or piece of dowel.

2. Complete the chart.

Figure	Shape of Faces	Shape of Base/Top	Name of Polyhedron
A			
B			
C			
D			

3. Create more polyhedra to add to your mobile.